LONGEVITY

Why we are living longer than ever and the discoveries that may allow us to live to 1000

'To die of old age is a rare, singular, and extraordinary death, and so much less natural than others: it is the last and extremest kind of dying …'

The Essays of Michel de Montaigne (*French philosopher, 1533-92*)

David Goldhill

CONTENTS

Acknowledgements

I wouldn't be here without my parents. They gave me life and passed on the millions of years of evolution embedded in their genes. Sadly my mother died all too young from a condition that, nowadays, she would probably survive. My stepmother celebrated her 90th birthday in style. She is a testament to the benefits in longevity that have accrued during my lifetime, and a rebuke to those who think that there is something undesirable about living a long time. I could not have written this book without the support of Lucinda, my wife. Many years ago she decided that I was worthy enough to make the most important decision any of us can make, and consented to merge our genes. They have been combined to give rise to three wonderful children who have, so far, created four grandchildren. To our children I would say you do not choose your parents but in the game of life you've been dealt a pretty good hand where your genes are concerned. It's now up to you to play the hand as best as you can.

We should all acknowledge a debt to the scientists who continue to grope towards a clearer understanding of our biology and the universe we inhabit. The knowledge in this book has only come about because someone has spent hours peering through a microscope at a minute worm or days waiting for a fruit fly to do something interesting.

I am also grateful for all of those who have tolerated my outpourings about longevity, and those who have helped me with the science and pointed out where I had made errors. Among these is Carina Kern who reviewed the accuracy of some of the statements included in the book. Michael Crozier deserves a special mention. He has turned my writing and ideas into a coherent and readable form.

There will be mistakes in this book. I do not pretend to be an expert in the complex and fast changing science of ageing. I plead guilty to gross oversimplification. I have neither the expertise nor the desire to produce a book of biology or genetics. If you're looking for the secret to unlock a thousand year life you will not find it here.

But then no one else can tell you because no one knows. I will have left out or skated over areas that will later be shown to be central and may well have placed emphasis on matters that, in time, will prove to be of little consequence. My intention is to arouse your interest, provide a few interesting facts and to suggest some ways that progress towards increased longevity may be made. Any errors are mine alone. I have enormous confidence in today's generations and those to come. I only regret I will not be around to see the world they will create.

David Goldhill, London, 2018

Why we are living longer than ever, and the discoveries that may allow us to live to 1000.

Introduction

Death has been a constant and familiar companion to all living creatures that have ever existed on Earth. For humankind, life has often been short and, until recently, few people achieved their full potential lifespan to die of old age. During the last 150 years the chance of an individual surviving childhood and achieving something close to their natural lifespan has increased enormously. This transformation has affected all people throughout the world, wherever they live and whatever their social standing and income. It is an inspirational story about turning scientific and medical knowledge into practice. Looking ahead developments in medical science may allow people to live extraordinarily long lives, well beyond what is achievable without intervention, and much longer than anyone has ever lived before.

We don't have to go back more than a few generations to a time when death was a familiar event. Death happened at home, in the fields, on the battlefield, on ships or in the workplace. Large families were the norm and many children did not survive to reach adulthood. Giving birth and being born were both dangerous, with high infant and maternal mortality. Hospitals, where they existed, had little to offer, and were not the place where most people died. Death was capricious, visiting the young as well as the old, the hale as well as the infirm, the rich as well as the poor. It often came with little warning killing individuals, families and whole communities. It was an accepted and everyday occurrence over which man had little influence. No wonder that it was often seen as beyond man's control.

I'm now retired but I spent many years looking after seriously ill patients admitted to an intensive care unit. Most of these patients were like prisoners on death row, certain to die without intervention. For some there was no last minute reprieve. But for many we managed to keep the grim reaper at bay, at least for a little while. I worked as part of a team of doctors, nurses, therapists and technicians. We needed equipment to diagnose, investigate and support our critically ill patients. We had access to a pharmacopoeia of drugs to treat infection and maintain critical organs. We needed our skills, knowledge, technology and drugs to save these patients. During the last 100 years or more, medical science has been remarkably successful in tackling disease and allowing most people to reach their potential lifespan. Because of these advancements millions of people are alive today who would have died not so many years ago.

Medicine, I believe, stands on the brink of a revolution in understanding and therapeutic options as great as anything that has previously been seen. The focus is moving away from treating problems when they arise to anticipating and preventing disease and deterioration. In the same way that modern aircraft have a maintenance schedule and programmed replacement of key parts, it is likely that worn-out organs will be overhauled, repaired or replaced before catastrophic failure intervenes. More radically, and in the same way that aircraft design has developed in a mere 100 years from flimsy wood and wires to gigantic metal mass people transporters, jet engines, rockets and hypersonic speeds, so the tantalising and somewhat scary possibility of redesigning *Homo sapiens* is something to be considered.

In recent years the search for ways to achieve an extended lifespan has moved out of the realm of fantasy and science fiction to become a legitimate scientific pursuit. Death has moved from being inevitable to a technical hitch, something amenable to intervention and prevention.

Many credible investigators in reputable institutions are looking into the causes of ageing and methods to slow, stop or even reverse it. Several wealthy individuals have donated large sums of money to fund such research. Companies with names such as Human Longevity Inc. and Elysium Health have been set up to take advantage of the progress that is being made, and even the internet giant, Google, has got in on the act by establishing Calico, a biotech company with the goal of combating ageing. If you believe, as I do, that this is your one and only shot at life, I quite like the idea of prolonging the time I have down here before I shuffle off my mortal coil. Even better I'd like to be restored to a physiological age of about 30.

In this book I have tried to tell the story of longevity, the time each of us has been given to live our lives. No one can predict the years they have been allotted but it is only within the last few decades that most people with access to modern medicine can reasonably expect to survive to reach old age.

To set the scene I look back at what life used to be like hundreds of years ago. Some of the events that have occurred since my grandfather was born illustrate how much has been discovered and how these discoveries have altered how long we live. I've tried to show how fast change is happening and the reasons why we can anticipate even more innovation in the future. I detail what goes into making a human body, explore the miracle of life, the meaning of death and some of the reasons we get old. I look at the changes that have already transformed life expectancy for millions of people. More and more people are now able to achieve their potential lifespan and I examine some of the ways that this can be realised. Modern medicine has become much better at preserving and repairing the organs that are essential for life. I extrapolate from current developments to imagine what interventions may be feasible in the near future.

Already many individuals are living well over 100. Some commentators believe that much longer lives are a possibility. A self-described biomedical gerontologist, Aubrey de Grey, has put his cards on the table and said that the first person to live to be 1,000 years old is alive today. It may sound far-fetched but there are good reasons to believe he just might be right.

Living for hundreds of years is going to require much more than tinkering with the protoplasm we have been born with. I cast an eye over some of the ways in which this may be achieved. Someone who lives to be 100 is a centenarian. A millennium is 1,000 years. So finally I conjure up a world populated with millenarians and give some personal speculations what it might look like. I do this with the understanding that the future will almost certainly be nothing like the one I imagine. The recent past has managed to confound and surpass the predictions of futurologists. I expect nothing less of what is to come.

My career in medicine taught me that people sometimes uncritically believe things that they read or someone tells them. We once had a patient in our intensive care unit whose family believed in miracles. They wanted us to keep him on life support hoping for a miracle to cure his catastrophic brain injury. The miracle that would have happened if he had survived being taken off life support wouldn't do. I have friends who believe in reincarnation, astrology and the medical benefits of homeopathy. Some may say people are gullible to believe the incredible and unsubstantiated. When it comes to the struggle between life and death I prefer to acknowledge that those in desperate straits like to cling to hope and grapple to understand even when no explanation exists.

Personally I prefer there to be concrete evidence before making up my mind. The human body is an unbelievably complex, dynamic and imperfect organism. Trying to understand fully our biology is an heroic undertaking, and it is even more difficult to obtain proof about the effectiveness of many medical interventions designed to extend life. My career in critical care medicine was notable for the large numbers of treatments that we thought were useful but were later shown not to work. Despite all this, outcomes for patients improved markedly over my working life.

In this book I have cast a sceptical eye over some of the evidence. Harley Street in London is full of qualified doctors, some of whom are snake oil merchants, willing to sell their patients what the patient wants, not necessarily what they need or what works. Where money is to be made you will find someone willing to sell you a dream, even if the facts don't stack up. If you want to spend your money on realigning energies or sitting under a curative crystal then this book is probably not for you.

I have done my best to stick to the facts where they are known, and to interpret and interpolate them in a way that is not too fanciful. I am taking an optimistic view though, in the hope that medical science can continue to understand our biology and make further progress in helping us to live long and healthy lives.

I have carried out my evolutionary task by contributing my genes to three daughters who have, to date, spread my portion on to four grandchildren. There are relatively few years left to me so why spend precious hours reading and writing about this subject? I feel extraordinarily lucky to live in our present age. Not only does my generation have a good chance of living longer than any previous one, but I have lived through the most remarkable period of scientific endeavour. It has been an incredibly exciting and invigorating time to be alive.

Mankind's curiosity has taken us to the stage when we can start to answer the fundamental questions about ourselves, why we are here and what are the mechanisms that got us to this point. I believe that we stand on the threshold where this understanding can be translated into the most extraordinary and wonderful changes in the human condition. Obviously, there are dangers and challenges from both man-made and natural forces. However, looking forward with optimism the world could be transformed into a healthier, happier and better place. I can only regret that I am not young enough to be around to see many of the remarkable developments that I think will take place over the next 50 to 100 years.

I hope to communicate a little bit of the excitement and wonder that I feel when reviewing some of what has been learned over my lifetime. If this message inspires a single person to contribute to adding to our knowledge then I would have done a little bit more for my species than just passing on my genes.

1 Shaping our own destiny

"It is not in the stars to hold our destiny but in ourselves."
William Shakespeare. (This is a common misquote from *Julius Caesar* [Act I, Scene II]. In it Cassius actually says *"The fault, dear Brutus, is not in our stars, but in ourselves, that we are underlings."*)

The ancestors of *H. Sapiens* have been around for about six million years, although our species first appeared some 200,000 years ago. Agriculture dates back 10,000 years, and the Bronze Age of about 6,000 years ago heralded the end of the Stone Age and the beginnings of civilisation. Carl Sagan popularised the cosmic calendar visualising the history of the universe as a single year: The Big Bang took place some 13.8 billion years ago. This is assumed to be at the beginning of January 1 on our calendar and the present day is midnight on December 31. On this calendar, first life on Earth formed towards the end of September. Multicellular life took until early December to evolve. The dinosaurs flourished and became extinct in late December. Apes evolved early in the day on December 31 and primitive man with stone tools appeared at 22:24 that evening. The early Bronze Age was 13 seconds before midnight, Jesus was born five seconds before midnight and Christopher Columbus three seconds later.

It is extraordinary that our species has evolved so quickly to dominate our world. We have thin, vulnerable skin, and no claws, horns or antlers with which to attack or protect ourselves. Evolution is not a progression towards perfection but it is a matter of best adapting to the environment. Many species before us have existed, evolved and become extinct. We are not the pinnacle of evolution but simply one of many adaptations. However, we are unique in being on the threshold of understanding our own biology.

If our civilisation survives a few more seconds on the cosmic calendar we are likely to be able, through our own actions, to alter fundamentally who we are and how long we can live.

Looking back a mere 1,000 years to the Middle Ages demonstrates the pace of change that characterises our era. In 1016, Ethelred the Unready died and King Canute II assumed the crown of England. Total world population is estimated to have been about 300 million. Much of the world was unknown. The fastest anyone could travel on land was on horseback. Several centuries would pass before the Renaissance and then the start of the Scientific Revolution. It would be nearly 500 years before Christopher Columbus discovered the Americas while sailing westward in an attempt to reach the East Indies.

The microscope was invented in about 1590. The human eye can just about resolve two lines 0.03 millimetres (mm) apart. The microscope led to the discovery of a whole new unsuspected world teeming with life, a world of microorganisms that exists under our noses, as well as in them, and in every conceivable orifice, cavity, pond or surface (a typical bacterium is about 0.2μm across and 2 to 8μm long – one μm is a one millionth of a metre – in other words very, very, very, small). The microscope also allowed us to look inside ourselves and describe the cell, the basic unit of which all life is constructed. The telescope first appeared in the early 1600s leading to a complete reassessment of the place of our planet, and of mankind itself, in the Universe.

I remember my grandfather well. I was 26 when he died aged 84. He was born in 1894, one of seven siblings as was common back then. Queen Victoria had been on the throne for 57 years and had another seven to go. Only eight years earlier Karl Benz invented the first petrol-powered car. Parts of the world were still totally unknown. It would be 14 years (or possibly 15 – the claims are disputed) before anyone reached the North Pole and it was not until 1911 that Amundsen reached the South Pole.

In those days childbirth was hazardous for both mother and infant. Life expectancy at birth was much lower than it is today. Death was common from diseases that have now been conquered or rendered far less deadly with immunisation, antibiotics, sanitation, better hygiene and nutrition. My grandfather was fortunate to survive the First World War that devastated a generation of young men. The world then was a far more dangerous place. Without modern medicine, blood component therapy, intensive care and countless other innovations and interventions, many died who could now be saved. When my grandfather was born the population of the world is estimated to have been 1.6 billion. By 1920, the year of my father's birth, it had risen to 1.9 billion despite the horrors and carnage of the First World War and the flu pandemic that killed millions as the war was drawing to a close in 1918. Between those years X-rays were discovered (1895), Max Planck formulated the basis for quantum theory (1900) and Albert Einstein published his theory on general relativity (1915). The Wright brothers in 1903 made their short, tentative flight at Kitty Hawk (The eminent physicist, Lord Kelvin, is famously reputed to have said in 1895 that heavier than air flying machines are impossible).

I came along in 1952 by which time World War II had been and gone and the population of the world was some 2.6 billion. In the intervening period many important inventions had occurred, including penicillin (1928), the jet engine (1930), the atomic bomb (1945) and the transistor (1947). In 1936 the BBC started transmitting television signals in black and white.

The year after my birth Crick and Watson identified the molecular structure of DNA (deoxyribonucleic acid) and Hillary and Tenzing reached the summit of Mount Everest. The invention of the microprocessor (1971) has changed the world forever, and we are in the infancy of the transformation that computers and artificial intelligence will bring. Stem cells were first recognised in human cord blood in 1978 and they are now used to treat many medical conditions with the promise of many more therapeutic uses in the future.

When my daughter was born in 1980 there were 4.6 billion people sharing the planet with her. The Berlin Wall fell in 1989 and changed the political geography of Europe. The first mammal cloned from an adult somatic cell was a sheep named Dolly born in 1996, thus demonstrating that a non-reproductive cell could develop into a whole individual. As recently as 2003 the first sequence of an individual human genome was published.

By the time my daughter's daughter, my granddaughter, was born in 2016 the population of the world had swelled to 7.4 billion. The world my granddaughter was born into would be unrecognisable to the horse-drawn, steam-powered world of my grandfather's birth.

There is no part of the planet's surface that is unknown and that cannot be viewed from an aircraft or from space or by browsing Google Earth.

The pace of change has been remarkable. Never have there been so many people on planet Earth. Also so many of them are highly educated with the skills and tools to find out more about ourselves, our world, and the universe beyond. Scientific method has been remarkably successful in advancing knowledge in a way that no amount of contemplation can achieve. The number of scientists doubles roughly every 18 years so it should not be a surprise that what we know is increasing at an exponential rate.

My grandchild can reasonably expect to live well into the 22nd century. But that is without accounting for medical progress. I stand as the link between my grandfather and my granddaughter. What will her world look like, and that of her grandchildren? Maybe she will be the first person to live to be 1,000. I celebrated the coming of second millennia. Perhaps she will be around for the third?

The past gives us few clues as to what will happen during the next 1,000 years. The baseball player Yogi Bera once said: "The future's not what it used to be." He's right, we live in unprecedented times and the pace of change and innovation is explosive. Never before in human history have we understood so much and stand on the brink of knowing so much more. Could the subjects of King Canute have imagined our era, and how much more difficult it is for us to look 1,000 years into the future? We are on the threshold of an incredible flowering of understanding and applying this knowledge about the way our bodies work, and about the way our universe is governed by natural laws.

The scientific method has been extraordinarily successful in adding to our knowledge and applying it to our world and to ourselves. Every indication is that the pace of progress is accelerating.

Unlike any other previous era, mankind now has the means to destroy the world so completely and in so many ways. Nuclear holocaust, global warming, disease, famine and social unrest all have the potential to demolish civilisation. Natural disasters such as volcanic eruptions may plunge us into a nuclear winter, or a comet may wipe us out like the one that is thought to have ended the reign of the dinosaurs.

If mankind does not succumb to manmade or natural disasters, and continues to progress exponentially, it becomes difficult to predict the next 100 years, let alone peer darkly into the distant future. Looking back can only make us realise how utterly different and unimaginable the world will be 1,000 years from now.

Let's be optimistic. Let us assume that the world continues on its stuttering but ultimately upward path. There are more people alive than ever, more have been clothed, fed and educated, and many more are involved in pushing at the doors to acquire and apply even more knowledge. We know that experimentation, questioning and methodical approaches to solving problems have a tremendous success rate at teasing out the truth.

In medicine I would suggest that three approaches, all in their infancy, have the potential to revolutionise treatment and may all have their place in prolonging human life. These are stem cell therapy, genetic manipulation and the application of technology to replace and enhance human functions. I write more about each of them later.

There is often resistance to new ideas and a failure to accept hard evidence. Standing in the middle of an African plain it is easy to believe that the world is flat. If you watch a ship sail out to sea you may think it has gone over the edge as it disappears beyond the horizon. You would have to think again when the ship returns and reports nothing except sailing without any problems beyond the horizon and back. Some fairly simple trigonometry is all that is needed to obtain a reasonably good calculation of the Earth's radius and thus its circumference. Even the "Truth" may only be an approximation. It is enough for most people to know that the Earth is a sphere. Except it's not. The surface is far from smooth with numerous peaks and troughs from Mount Everest at 8,848 metres above sea level, to the depths of the Mariana Trench 11,034 metres below sea level. Furthermore the world is squashed at the poles and swollen at the equator so that the distance from the centre to sea level is some 21 kilometres greater at the equator than the poles. There are even subtle changes caused by the changing weight of the atmosphere and oceans as well as the effect of spin. All these findings merely refine the fundamental concept. Knowledge that the world is a globe spinning in space is the fact that changes everything.

New knowledge can make us completely alter the way we see the world. It's been a long time since any rational person thought the Earth was anything other than more or less spherical (although an internet search for the Flat Earth Society will provide you with an alternative viewpoint). It no longer seems far-fetched that we don't fall off a curved surface that, at the equator, is spinning at more than 1,000 miles per hour while orbiting the Sun at 66,000 miles per hour. The Sun is one of more than several hundred billion stars in our galaxy and our solar system is moving at some 43,000 miles per hour roughly in the direction of the constellation of Lyra. The entire Galaxy is spinning and our planet is rushing around the Milky Way at about 483,000 miles per hour. Our galaxy is one of an estimated 100 to 200 billion galaxies in the Universe. In the early 1600s when Galileo Galilei first stared up at the Sun, moon and stars, he could not have had any idea what the telescope would reveal.

In 1687 Sir Isaac Newton propounded his theory of gravity. This was a revolutionary insight not least because it followed that all objects are accelerated at the same rate in a gravity field, overturning the possible more intuitive belief that heavier objects fall faster than lighter ones. (So, in the absence of air resistance, a 1,000kg baby elephant will fall with the same acceleration as a monster 1kg mouse)

Newton's Theory explained a great deal, not least when it was used to predict the existence of Neptune based upon the wayward motion of Uranus. When a discrepancy in Mercury's orbit could not be explained by Newtonian theory the scientists had to go back to the drawing board.

The world had to wait until 1915 and Einstein's Theory of General Relativity to have an explanation for the finding. Not that this is the end of the matter. As I write eager academics are attempting to pin down dark energy, dark matter and multidimensional time and space. In the meantime, most people are happy enough to grasp that we have a pretty good theory of gravitation but are wise enough to understand that it may not be the final and complete explanation.

The infinite monkey theorem postulates that a monkey hitting typewriter keys at random for an infinite amount of time will, by luck, type out the complete works of Shakespeare. However, the wisest of men can sit for eternity round a body speculating about its biology and they will never reach the right answer. For much of human existence, disease has been blamed on spiritual causes, the whims of gods and the effects of spirits and humours. The dissection of a dead body was usually forbidden, mainly because of a belief in the afterlife and the fear of compromising existence there. Animals could be examined though, and ideas developed from physiological observation.

The question of the seat of the soul was widely debated among pre-Socratic philosophers. Aristotle placed the soul in the heart attributing to it emotions, whereas the function of the brain was to cool down the warm blood coming from the heart. Egyptian physicians believed that a man has eight substances including his name and his shadow. For some reason Hippocrates placed intellectual functions in the left cardiac ventricle, the largest of the four chambers in the heart. In the 3rd century BC, Greek anatomists were, for a while, allowed to dissect the body, but were still seeking a location in which to place the soul.

The Greek physician Galen lived from 129 to about 199 AD. His theories dominated Western medical science for more than 1,300 years. He believed that the workings of the human body were based upon four humours – blood, phlegm, black bile and yellow bile. These in turn were associated with the fundamental elements of air, water, earth and fire. Disease was the result of a bad balance between the humours. He also established the concept of the triple spirit – the animal spirit in the heart, the physical spirit in the liver and the psychic spirit in the brain.

Some traditional forms of treatment survived for many centuries. Bloodletting was one of the most common and probably caused enormous harm over the centuries. Cupping is another common traditional therapy. Some believe cupping acts by correcting imbalances in the "internal bio field", such as restoring the flow of "Qi" (I have no idea what this means). Cupping is still discussed in glossy magazines as well as reputedly being used by various celebrities.

Centuries of pointless theorising were eventually replaced with practical experimentation. In 1628 the physician William Harvey described the circulation of the blood. He came to his conclusions based on dissection with close observation. As always there were those who found a reason why experimentation should not be used. How, they argued, could you find the general laws of nature by creating highly artificial situations in an experiment? It was not until the 1800s that the elements of modern medical science started to develop. These included clinical research, laboratories, government funding, professional associations and scientific journals.

It doesn't matter how passionately or how widely held a belief may be. Throughout history many "self-evident" facts have been shown, in time, to be wrong. It requires an open and enquiring mind and a willingness to reach conclusions based upon what is seen and not upon what one would like to happen. Scientific investigation has been remarkably effective in helping us understand our biology. However, there is little doubt that some of what is now scientific orthodoxy will be shown to be wrong, and many other facts will be modified or qualified.

When interpreting medical evidence several things need to be borne in mind. A single case report rarely provides definitive proof that something works. If a friend's back pain improves after a visit to an osteopath that does not mean that the treatment will work for you. Osteopathy may have been effective in this single case or perhaps the chronic back pain was spontaneously improving. Much more hard evidence is needed to prove that it works most of the time for most people.

Another common problem is to confuse association with causality. Just because two measurements relate closely to each other does not mean that one is the cause of the other. Finding an association is often important as it may lead to further investigations uncovering evidence of causality. The numbers who drown by falling into a pool may correlate with the number of films Nicolas Cage has appeared in. It is pointless looking any further into this association. On the other hand, the association between smoking and lung cancer was important evidence linking tobacco to harmful effects. Another pitfall is to assume that a secondary or proxy outcome means that a primary effect will follow. For example, many foods have potential benefits as antioxidants or cholesterol-lowering agents. This does not mean that they will have any measurable impact on overall health or mortality.

Normally, a randomised controlled study is the best way to get medical evidence. Basically, this involves randomly assigning subjects to get an active treatment or an inactive placebo. Predetermined outcomes are then recorded and compared. Both the subjects and researchers should be in the dark about the treatment. This is known as a double blind study. Careful statistical analysis is applied to see whether the results could have arisen by chance. Typically when there is less than a 5 per cent chance of this happening the treatment is taken to have been effective. This still means that the positive result could have happen by chance once in every 20 times. The results are only applicable to the patients studied and caution must be used when applying the findings to a much bigger and wider group of people. Just because the treatment is shown to be effective on statistical analysis in some selected subjects does not mean it will be effective in all of them or would be in others of different ages, races or genetic variations. This sort of study looks at simple outcomes related to the expected effect. It may not pick up subtle or long-term harm that becomes apparent only after careful analysis of long term use in many patients.

Getting to the truth is far from easy.

Although unbiased data collection and analysis is essential to be able to reach robust conclusions, formal research is not always necessary. People have leaped from flying airplanes without a parachute and survived. Others have jumped with a parachute and died. Do we need to do a study to demonstrate that parachutes are effective at saving life when leaving an aircraft at altitude? For obvious reasons, it will be impossible getting rational volunteers to take part in such a study. Research is expensive, complicated and time-consuming. It often gets reinterpreted, revised or rejected as more evidence accumulates.

Despite its recognised limitations scientific research has been remarkable successful in helping us understand ourselves and the world we inhabit. This knowledge has been translated into practical solutions that have fundamentally changed who we are, how we live and how long we live.

As a species we have made great strides towards understanding the universe we inhabit. We will discover much, much more about it in the future. Perhaps there are some things that will forever be beyond our comprehension even with artificial intelligence to help.

I believe it is necessary to acknowledge and accept that there are things we do not know or fully understand. This must be the first step towards searching for the answer.

As former United States Secretary of Defence, Donald Rumsfeld said: "… *there are known knowns; the things that we know. We also know there are known unknowns; that is to say we know there are some things we do not know. But there are also unknown unknowns – the ones we don't know we don't know*". Unforeseeable events can play havoc with any prediction. Looking ahead, known unknowns could be practical cheap fusion power, hydrogen-powered cars, meteorite mining or alien contact. It will be interesting to see what unknown unknowns appear.

I can be sure that the future will be very different from our present. Each generation has concerns about their children. My parents were worried about long hair, pop music and loose morals. Despite their fears I managed to study, have a good job and help raise a family. My children and grandchildren will be different, have different passions and attitudes. That's our nature. Giving women the vote, letting gay men and lesbians marry and allowing children to speak and be heard, hasn't meant the end of civilisation. Mankind has adapted to much greater events. We manage to accept that the Earth is not the centre of

Timeline of events and world population growth

1798	Malthus on population growth
1804	**world population one billion**
1895	X rays discovered
1903	Wright brothers heavier than air flight
1909	North Pole reached
1911	South Pole reached
1914	WW1 starts
1915	Einstein publishes theory of general relativity
1927	**world population two billion**
1928	penicillin discovered
1936	BBC television broadcasts
1939	jet engine powered flight
1939	WW2 starts
1945	atom bomb invented
1947	transistor invented
1953	DNA described
1953	Everest conquered
1959	**world population three billion**
1961	genetic code described
1969	moon landing
1971	microprocessor invented
1974	**world population four billion**
1978	stem cells identified
1987	**world population five billion**
1989	fall of the Berlin Wall
1996	mammal cloned from adult somatic cell
1999	**world population six billion**
2003	human genome described
2011	**world population seven billion**

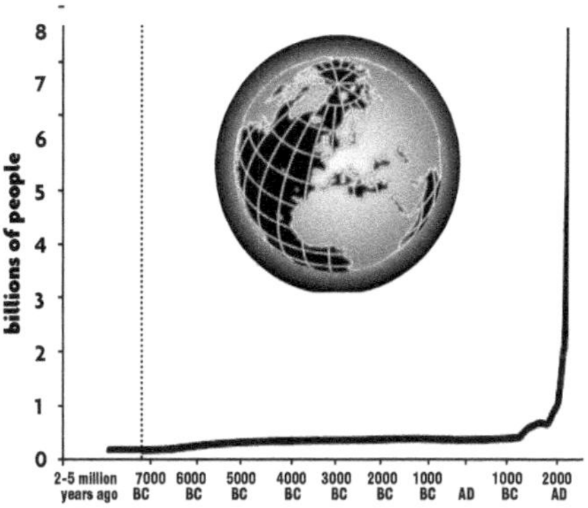

the Universe, *H. sapiens* is not the pinnacle of creation, and God is not essential to account for us and our world. Our species is adaptable. I think it is our defining characteristic. We accept as rational getting old and dying. Perhaps in time that will be seen to be as ridiculous as denying pain relief to those who are suffering.

If we don't destroy ourselves I have every confidence that our species will adapt to living long and very different lives. Will there be 1,000-year-old men and women? I wouldn't bet against it.

2 Building blocks

"A man is a god in ruins."
Ralph Waldo Emerson (*Nature*, 1936)

Every human that ever lived was made from a handful of basic building blocks forged in the maelstrom of a star and then flung out into space. Only the lightest elements, hydrogen and helium, were created in the Big Bang. The other elements were made inside stars. The elements of which we are formed ended up on planet Earth. Over millennia elements formed more complex compounds. Somewhere the right chemical mix and environment combined for life to start. We still don't know whether this was an exceptional and rare event, or whether it is common throughout the Universe. Either way life is extraordinary. On our planet life has evolved into millions of different forms filling almost every conceivable habitat. However, I also find it extraordinary that all life on Earth shares common characteristics and ancestry.

The periodic table describes the range of elements, the unique building blocks of which our universe, our world and ourselves are made. Elements cannot be changed or destroyed or turned into simpler substances with chemical reactions. The first 94, from hydrogen to plutonium, occur naturally, although some are only found in trace amounts. The number indicates the number of protons within each atom. To date elements with 95 to 118 protons do not exist in nature but have been synthesised. Each element is itself made up of sub-atomic particles.

The first to be described was the electron in 1897. Physicist Ernest Rutherford is credited with the discovery of the proton in 1911, and identification of the neutron took place in 1932. Since then discoveries have become much more complicated with experts counting anything from 17 to 37 fundamental particles depending on spin, charge and mass, not forgetting flavour or colour as well as anti-particles. I will leave it to the physicists to try and make sense of all of this.

For life to exist simple elements have to combine into more complex compounds. These are exactly the same elements that make non-living substances. All Earth-based life combines a small number of elements to construct a bewildering array of organic molecules to perform the complex chemistry that is essential for life.

Of all the elements within the body the key one is carbon. Although only the second most common element by mass in a human body, this versatile substance has unique chemical properties that make it ideally suited to its central role in building a body. It is able to link with many other elements to form large numbers of different organic compounds. It is something of a miracle that carbon is so abundant. A precise synthesis of this element is needed to enable carbon-based life. It may be an amazing coincidence that this state exists. On the other hand if it didn't then none of us would be here. For those who believe in the possibility of alternate universes, and many reputable physicists do, then other worlds may have different laws of physics. There may be planets where carbon is scarce or where it is common but relies on other reactions for its formation.

The most common element in a body is oxygen, which combined with the third commonest element, hydrogen, makes the water (H_2O) that accounts for some 65 per cent of the mass of each of us. The three elements, oxygen, carbon and hydrogen, make up about 93 per cent of our bodies. Nitrogen contributes another 3 per cent, calcium 1.5 per cent and phosphorus 1.2 per cent. The rest is made up potassium, sulphur, sodium, chlorine and magnesium, and a handful of trace elements. Some of these trace elements are essential to life whereas others are contaminants. Even elements essential to life may be toxic in higher amounts.

All the elements that are used to make us are neither created, nor

Elements in a typical 70 kg adult human

	%	Weight (kgs)
Oxygen	65	45.5
Carbon	18	12.6
Hydrogen	10	7.0
Nitrogen	3	2.1
Calcium	1.5	1.05
Phosphorus	1.2	0.84
Potassium	0.3	0.21
Sulphur	0.2	0.14
Sodium	0.2	0.14
Chlorine	0.2	0.14
Magnesium	0.05	0.04
Trace elements	‹1.	

Boron, chromium, cobalt, copper, fluoride, iodine, iron, manganese, molybdenum, selenium, silicon, tin, vanadium, zinc

are they destroyed when we die. They carry on available to be used again. Thus an atom of carbon could exist in pure carbon compounds such as graphite, coal or a gleaming diamond, all 100 per cent carbon and each extraordinarily different in their form, properties and worth. The carbon atom could be incorporated into substances such as petrol or methane gas, be in plastic, alloyed with iron to form steel or be in the atmosphere as carbon dioxide or carbon monoxide. It may return to be part of a living organism in a bacterium, plant, insect, fish or human rocket scientist.

Almost all living creatures are built using a small number of carbon-based molecules. They consist of three simple sugars, 20 amino acids, five nucleotide bases, six lipids and a solitary phosphate group. We know what elements and molecules go into building a living human. We know that these elements combine to make more complex molecules which interact to form tissues and membranes, which organise to make organs such as the heart, the kidneys and the liver, which produce enzymes and hormones and neurotransmitters that control everything from our digestion to our moods.

The magic of DNA in each cell holds the blueprint for our construction as an individual distinct and different from all others, except for identical twins.

We take in fuel in the form of food and process it to provide energy and materials to build our bones and muscles and blood vessels and brain and everything else that makes the human body.

Living things are different from inanimate objects as they obtain energy from their environment, perform chemical reactions, respond to their surroundings, change and develop, reproduce and share a common evolutionary history. All living organisms use the same chemical building blocks and the same cellular organisation. This applies whether that life is a minute bacterium or a giant redwood tree. Perhaps this is the only form that life can take. Or perhaps other forms of life came into being and didn't manage to survive and evolve. It is therefore believed that all forms of life that currently exist on Earth are descended from a common ancestor that lived about 3.5 billion years ago. That means that if we trace any living thing back through its ancestors we will eventually get to the same source, the Universal Common Ancestor. All living things are organised into cells and the creation of the cell is fundamental to life. Here, things can get a bit complicated. A virus doesn't have a cell wall but rather a protein coat. Is it life? It can't do much for itself and hijacks a cell's machinery to carry out all its functions. As for a prion; can a misfolded piece of protein be life? A cell simply consists of a wall or membrane that encloses contents. This means that what is inside can be different from what is outside. The inside of a cell can therefore be organised and protected.

Single cells eventually learned to cooperate and then to specialise. The multicellular organism was born.

3 DNA: The book of life

"If life must not be taken too seriously, then so neither must death."
Samuel Butler (*The Note-Books of Samuel Butler*, 1912)

It was not until the mid-1800s that microscopes capable of accurately observing cells were available. The cell is the basic unit of life. There are only two basic types of cell; the **eukaryotic** and the **prokaryotic**. Eukaryotic cells are those which animals, including humans, are made of. Within eukaryotic cells is another membrane containing the nucleus. This is where the genetic material, or chromosomal DNA (deoxyribonucleic acid), is contained. Other membranes enclose specialised organelles (organised structures within a cell) that, in humans, include mitochondria that are the energy-generating factories, as well as the molecular machinery necessary to interpret and protect genetic material, replicate itself and carry out its instructions. By contrast prokaryotic cells do not have nuclei or membrane-enclosed organelles. All bacteria are prokaryotic cells and consist of a single cell.

The nucleus of the cell usually makes up some 5 per cent to 10 per cent of the volume of the cell and contains all the genetic material in the form of DNA. With a few exceptions, such as the red blood cell, human eggs and sperm, each and every cell contains the full set of DNA sufficient to reproduce every other cell within the body. I am not a geneticist nor is this a textbook on genetics. However, a little more understanding will be useful in explaining the role that genes play in longevity, and the ways in which genetic manipulation may be important in the future in terms of longer, and healthier, life.

As recently as 1944 DNA was clearly identified as the place where genetic material was located. Nine years later James Watson and Francis Crick described the double helix structure of DNA. How they came to make the discovery is a fascinating story worth reading about.

DNA consists of a string-like backbone, a double helix, made up of sugars and phosphates from which four bases protrude.

These bases pair up with each other to form a base pair, one on each of the helixes. The bases are adenine, guanine, cytosine and thymine (abbreviated as A, G, C, T). The purine A always pairs with the pyrimidine T (A with T) and the pyrimidine C always pairs with the purine G (C with G). RNA (ribonucleic acid) is similar in structure except uracil (U) replaces thymine. Its main role is to carry instructions from the DNA to the place in the cell where proteins are manufactured.

Finally, in 1961 Gobind Khorana and Marshall Nirenberg cracked the genetic code for protein synthesis. The breakthrough was to produce synthetic messenger RNA that contained only uracils. This yielded a protein containing only phenylalanine demonstrating that UUU was the RNA code for phenylalanine. **We now know that three bases in a specific order code for a specific amino acid**. The three base unit is called a codon. So, for example, ACA codes for threonine (as do ACT, ACC and ACG) whereas CAC and CAT specify histidine. A protein is a chain of amino acids and there are only twenty amino acids that combine to make any protein. The number and sequence of amino acids builds a unique protein and the proteins carry out the necessary biological work. Within a few years the codons for all 20 amino acids had been elucidated.

All species are based upon the same code and use the same molecular machinery. That is why it is said that we share 50 per cent of our genetic material with a banana plant.

A **gene** is the basic unit of inheritance passed on from parent to child and consists of DNA containing instructions to code proteins and to regulate and promote activity. A single gene can vary in size from a few hundred base pairs to more than two million. The human genome has about 20,000 genes, not that many more than a worm and a lot fewer than corn, rice or wheat. What matters is how the genes are used, not their size. Each gene has the DNA instructions to make proteins, on average three proteins per gene. Proteins have been described as nature's robots as they are designed to carry out every task the cell requires. Within the cells, ribosomes are the machinery that convert DNA instructions into proteins. New proteins are continually synthesised while old or damaged proteins are degraded and pumped out of the cell.

The genes only make up about two per cent of the DNA and there are large segments of DNA that do not contain any protein-coding information. The non-coding sections within genes are called introns and that between genes is intergenic DNA. The role of much of this non-coding DNA is unknown and may partly consist of "junk" DNA and legacy non-functional fragments unable to be eliminated or escape. Some is there to regulate when, how and how much of a protein is made and is therefore a vital component of genetic expression. The introns, may span hundreds of thousands of bases, and are the rule rather than the exception. The effect of a gene is commonly not on or off, black or white. They may have a range of effects (expressivity) or vary in the chances of having an effect (penetrance). Some genes may affect many traits. On the other hand, some traits are influenced by more than one gene.

Human DNA is packaged into chunks called **chromosomes**. Humans have 46 chromosomes in total. Half are inherited from one parent and half from the other. They consist of 22 matched pairs and either two X-chromosomes for a female, or an X- and a Y-chromosome for a male. At the end of each chromosome is a telomere about which more will be written later (see page 65). The total amount of DNA is called the genome and each genome contains more than three billion base pairs.

In 1977, Fred Sanger developed rapid DNA sequencing. Investigators have learned how DNA replicates and how synthesis is controlled. An understanding of the role of mutations has become clearer. A mutation is a permanent spontaneous change in the sequence of DNA. This can occur as a result of an error during DNA synthesis or from damage caused by radiation or certain chemicals. A mutation may occur at one or two nucleotides (the basic structural unit and building block for DNA) and consist of a base substitution, or an insertion or deletion of a base. Or it can affect large regions of the chromosome.

The first condition identified with a defined genetic abnormality was Huntington's Disease in 1983. 1983 was also the year polymerase chain reaction (PCR) technology was invented. This is a technique for making billions of copies of a sequence of DNA, an important step in the study and application of DNA sequencing to diseases, research and criminal investigation. Thousands of gene variants have now been linked to diseases or abnormalities. Some are caused by a single genetic mutation. A mutation in a single gene can affect many organs. An example is Marfan's, an autosomal dominant disease where a gene mutation affects the manufacture of a protein essential for the production of connective tissue. Those who inherit it have abnormal joints, heart valves and blood vessels. Other genetic conditions, such as Down's syndrome, are caused by extra genes; in this case a whole chromosome with some 300 genes. Conversely some diseases such as diabetes, high blood pressure, depression, infertility and obesity, are affected by many genes. There is no single gene for high blood pressure but many genes affect the physiology of blood pressure control that may result in the same outcome, namely hypertension.

The Human Genome Project was launched in 1990 to map the whole human genome. Haemophilus influenza has the dubious honour of being the first bacterium to have its genome sequenced in 1995. The first mammal was a mouse decoded in 2002. The culmination of the project at a cost of some $2.7bn was the publication of the human genome in 2003. A monumental effort by state-supported and private initiatives had succeeded in one of the most important scientific discoveries that mankind has ever made. We now know there are slightly more than three billion base pairs that together provide the instructions to build a human being. In 2016 it cost about $1,000 to sequence a human genome, one measure of the enormous progress that has been made in this field.

All humans (except monozygotic twins or clones) are genetically different. However, we are also almost identical to each other. The genetic difference between individual humans is about 0.1 per cent. There is a difference of some 1.2 per cent in the genes we share between *H. sapiens* and chimpanzees. Taking into account the total genome including introns and intergenic DNA, the difference may be as great as five per cent. Mapping the human genome was an essential step along the way to understanding where genes are located and what they do. From there it is but a small step to think about removing, altering or replacing undesirable genes. Genes are the language of inheritance, describing everything we are as human beings.

The genome is the book that has already been written and implanted into each and every one of our cells. Once we can understand the language then it becomes possible to edit and amend the book, or write our own stories.

The way the genes of mother and father are expressed determines the characteristics of the child. The **genotype** is the unique genetic information needed to build the individual and can refer to anything from a single gene up to the entire genome. Every one of the trillions of cell in the human body holds this information.

However, how this information is used defines the organism created. The difference is caused by genes being programmed to turn on and off at certain stages of development. So all the cells in our individual human body contain identical genes but some cells will turn on the mechanism to construct a nerve cell while others will become liver cells.

The **phenotype** refers to the organism's physical and biological attributes and characteristics. In other words what they look like and how they behave. The phenotype is based upon the genotype but is also influenced by external triggers. Factors known to influence the phenotype include the chemistry of the egg, hormonal and nutritional influences in the womb and physical and social factors to which the individual is exposed. These are called **epigenetic** effects. These usually come into play during an organism's lifetime.

However, these changes can be passed onto the next generation if they occur in germ cells (germ cells are the cells that develop into sperm and eggs. They contain half the number of chromosomes of other cells and unite with a germ cell from the opposite sex to form a new individual). Chemical tags and markers such as methyl and histone can change gene regulation, a possible mechanism through which gene expression is altered through non-genetic effects. Starvation, vitamin deficiency and stress have been shown to alter methylation (addition of methyl groups to a DNA molecule) and may be the explanation for the increased susceptibility to various diseases seen in individuals who have been exposed to these stresses. Interestingly, epigenetic modifications can be reversible which makes them a suitable target for intervention to change outcomes. Methylation patterns are correlated with many human diseases including cancers, autoimmune disorders and neurological conditions.

Identical twins share exactly the same genotype but will behave differently, are unlikely to be completely identical in looks, and will suffer different fates such as developing cancer, degenerative diseases or other conditions, and even sexual orientation. Individual cells contain the same genome but epigenetic influences determine its differentiation into a liver or a skin cell and so on. Which genes are expressed depend on epigenetic factors. Epigenetic influences can be profound.

If DNA is the book of life then epigenetics determine which paragraphs and chapters will be read and taken account of.

For example, a Japanese fish, *Trimma okinawae*, can alter its sex depending on the characteristics of the other fish within its shoal. This involves extraordinary changes in gonads, genitals, brain and behaviour. A caterpillar and the butterfly into which it develops have an identical genome, as do a tadpole and its frog, but the juvenile form of both is so different from the adult that they could be completely different species. Queen bees may live for years but worker bees with identical genes have a life span measured in a few weeks. They look different and have different roles. The answer also lies in the degree of gene expression and is driven by nutritional differences while in the larval stage.

Epigenetic factors may also modify inflammation, cellular survival, protein synthesis and other factors that may have a bearing on longevity. Changes in chromatin (the combination of DNA and protein that make up the content of the cell nucleus) have been identified in old cells, changes that may be a response to environmental stress. It is therefore believed that epigenetic changes occur as a response to the ageing process but may also be a cause of ageing.

The genotype determines the physical characteristics of the phenotype. It will determine gender. It dictates height and bone structure and the colour of eyes, hair and skin. There is little doubt it has some relevance to intelligence. More controversially it may well influence behaviour, impulses, personality and temperament. The age at which people first have sexual intercourse is partly influenced by genes that act on the timing of physical maturity and contribute to differences in personality. By one estimate perhaps a quarter of the difference in age when people first have sex can be explained by genetic variants.

Other genetically influenced traits may even include such things as attitude to risk, potential for addiction to drugs and alcohol, impulsiveness, novelty and attention seeking, religious beliefs, political persuasion and sexual orientation. Research into the genetic basis of these traits has been conducted by comparing twins, identical and non-identical, who were reared together or had been separated at an early age and reared apart. Throughout the world more than 800,000 pairs of twins have been studied. This has provided some insights into the relative importance of genetic and environmental factors. Identical twins, even when reared apart, share many of the same characteristics. These results have received some criticism and qualification. What is equally as relevant is why identical twins brought up in the same circumstances may develop in completely different ways. The answer may still lie in the genes. A study in worms with identical genes showed that random fluctuations in gene expression lead to changes in gut development.

In other words what you are and who you are may be partly a matter of chance.

There is also another small piece of DNA buried in another part of the cell. This is in the **mitochondrion**. Mitochondria provide much of the power supply for the cell. Each mitochondrion has its own DNA. It's likely that this organelle started off as an independent bacterium that decided it was better off taking up residence within our cells than continuing with an independent existence outside. Humans, as well as all other eukaryotes, have benefitted from having a reliable, efficient energy source. This is clearly an enormous evolutionary advantage. Where would humans be without it?

The story of mitochondrial DNA is fascinating. The mitochondrial genome is tiny, consisting of about 16 thousand bases encoding 37 genes. These genes are only inherited from the mother. They are very stable over many generations only changing as mutations occur. They can be used to track back populations in an unbroken line to a so-called "Mitochondrial Eve". She lived some 150,000 years ago. Mitochondrial disorders can cause a variety of diseases transmitted from mother to child. These diseases include neuropathies, myopathies and endocrine problems. One solution to this is the three-parent child – a recent development in IVF. In this case the mitochondrial DNA is supplied by a healthy donor and the rest of the gene package is supplied by the mother and father as usual.

4 Cellular life

"Life is what happens to us while we are making other plans."
Allen Saunders (*Reader's Digest,* 1957)

You might think that the prokaryotes are not relevant to being human. However, all eukaryotic cells have evolved from the prokaryotic precursor. Bacteria are essential for human life today. If you've ever thought you were alone, think again. We share our body with trillions of microorganisms. Not only have they decided to make their home in and on our body, but their wellbeing and existence is essential to ours.

Within our intestine, up our nose and on our skin there are as many bacteria as we have cells in our body, and maybe more. That's in the order of 30 to 50 trillion bacteria (a trillion is 10^{12} or 1 followed by 12 zeros). Add in fungi, viruses and other microbes and it has been estimated that perhaps one per cent to three per cent of our weight is made up of these creatures. On average more than 10 million bacteria from around 1,000 varieties make their homes on each square centimetre of skin. Within the colon each gram of faeces contains some 10^{11} microorganisms. Each episode of defecation may excrete up to a third of the microorganism in our intestines. These fellow travellers depend on our existence for their lives. However, they do their bit in return by helping with our survival. The intestinal flora coats the gut wall providing a barrier to harmful organisms. Disruption of normal gut flora has been implicated as relevant to several intestinal disorders. Some clinicians believe that introducing healthy faecal bacteria with a "stool transplant" is a good treatment for harmful *Clostridium difficile* (C diff) infection.

Children born by Caesarean section don't get exposed to the same bacteria as normal vaginal deliveries. Some studies have suggested that this may make the baby more vulnerable to asthma, hay fever and obesity. One response has been to expose the newborn to its mother's vaginal flora shortly after delivery. Microorganisms also play a role in enhancing and training the body's immune system to be able to recognise and attack harmful organisms.

Some bacteria have enzymes that the body lacks and can break down hard-to-digest dietary fibre and other polysaccharides. Without these microorganisms in their guts, germ-free rodents have to eat nearly a third more calories to maintain their body weight. Vitamins are vital nutrients that humans are not able to make from other compounds. They must therefore be obtained by what we eat. Although they were originally called "vital amines" thus vitamine, the e was dropped when it was realised that they didn't all have an amine group in their chemistry.

Gut bacteria are also a significant source for a range of vitamins, particularly those from the B group and also vitamin K. There is also evidence to suggest that gut microbiota influence human brain development and function.

There is communication between the brain and the gut through nerves, hormones and changes to the immunological system. If you ever thought that someone's stomach was their main organ of thought, there is now evidence to say you might be right. Finally it appears that a small number of our human genes, perhaps around 40, are bacterial in origin.

Some of these figures are mind-boggling and are worth emphasising. You, me, all of us, have bodies made up of tens of trillion of cells. Each cell has developed from an original single cell and each cell knows its function and place in the body. The mathematics of replication is extraordinary. A single cell divides into two and each of those cells divide so there are now four cells. Those four cells divide again to form eight and so on. Within 48 divisions there are about 10^{15} (10 million billion) cells.

Every cell contains the information to build any other cell or replicate the complete human.

Every day, even on Sundays, we make tens of billions of cells while the same number die. Within each cell, and forming only a small part of it, lies the DNA consisting of more than three billion base pairs. Straighten the coiled-up DNA and it would stretch to about two metres but weigh about 6×10^{-12}g (6 trillionths of a gram). Our bodies are crawling with trillions of creatures each of which has their own DNA.

Once a single human cell is created no direction or outside intervention is required to build the complete human. Even more amazingly most of this information was unknown until fairly recently.

Eukaryotic creatures encompass **Protista**. These are mainly tiny single-celled organisms such as amoeba. Another kingdom of eukaryotes is the **Fungi** that include single celled organisms (e.g. yeast) and multi-cellular ones such as mushrooms. **All plants and animals are multi-cellular**.

There are thought to be about 250,000 species of plant and there are more than one million known species of animal. Most of the animals, about 800,000 species, are in fact insects. Another 200,000 are sponges, jellyfish, worms, spiders, snails or sea urchins. The rest are birds, mammals, fish, reptiles and amphibians. However, the total number of organisms may be much greater as many living organisms are yet to be named.

Although current life on Earth is diverse, bountiful and widespread, it is a mere fraction of the total number of species that have existed and become extinct over the four billion year history of life on Earth. In fact it has been estimated that fewer than one per cent of all species that have ever lived are alive today.

Before I discuss prolonging life it may be worth thinking about what life is. All of us began when a single human egg (ovum) was fertilised by a single sperm to form a single cell with the DNA instructions to make a complete human. Usually fertilisation takes place in the fallopian tubes. The fertilised egg is called a zygote. Both the egg and sperm contain DNA, the egg from the mother and the sperm from the father. This DNA is the genome and contains the blueprint necessary to build the creature, and contains characteristics inherited from both mother and father.

That egg divides, through a process called mitosis, into two cells. Those cells divide to form four and then divide again so there are now eight cells, and so on. Each division takes some 12 to 24 hours. Over the next week the clump of cells carries on dividing, starts to differentiate, and moves into the uterus where it implants into the wall of the womb. The placenta and umbilical cord is the connection between the growing embryo and mother providing oxygenated blood and nutrition and carrying away waste products. The next eight weeks are the embryonic period of rapid cell division and organ formation. This is when organs such as the brain, heart and lungs, as well as the arms and legs, begin to form.

At a very early stage every cell has the potential to become any type of cell. At some stage chemical signals and interactions between cells cause them to become differentiated into specific types. Sometime around the ninth week the embryo is called a foetus (or fetus in American). At this stage it is about 5cm long and still has a lot of growing to do. Its head is much the same size as the rest of the body. The arm is a rudimentary limb that still has to develop the 43 muscles, 29 bones and extensive nerve supply to be fully functional. The foetus grows rapidly and the mother starts feeling it move by 18 to 20 weeks. A foetus born after 23 or 24 weeks of gestation has a greater than 50 per cent chance of survival, although only with the support that can be provided in a modern neonatal intensive care unit. By 26 weeks of gestation the survival rate with medical support is in the order of 80 per cent to 90 per cent. It is only when the mother carries the child to 34 weeks or more that there is an excellent chance the baby will survive outside the womb with little or no medical attention.

At some stage the growing foetus is recognised as a being an individual with an identity and rights, in other words as someone's child. There doesn't seem to be universal agreement on the particular stage of development that has to be reached. Some would say that life starts when the sperm fertilises the egg. The Human Fertilisation and Embryology Act takes 14 days as the point after which an embryo cannot be kept for research purposes. After this time the nervous system develops and the embryo will continue as a sole individual as it can no longer split into twins.

In an everyday biological miracle, that single cell is the genesis for the billions of cells that form the human baby. The cells replicate, organise and specialise. Buds that in other animals will turn into trotters or fins or wings, mutate into limbs complete with muscles, tendons, joints, fingers or toes. Despite common mythology, the growing foetus does not develop gills or a tail. In a miracle of organised development triggered by genes being activated and suppressed, the skeleton and nervous tissue grow, limbs appear in the right place and orientation, and organs develop and perform their specialised functions.

It has been estimated that the complete body will have some 3.72 x 10^{13} cells of more than 200 different types. Some will end up as hair, others fingernails, or skin, or gut, or liver, or bone, or brain, or any of the organs and tissues that are essential for human life. Many cells will die having done their job or been found to be faulty in some way by the body's defence and refuse systems. Throughout life cells die and are replaced in a process of constant dynamic renewal. How long they last depends primarily upon their function.

Studies on cell turnover use the spread of radioactive Carbon-14 worldwide following above ground nuclear testing. Plants, eaten by animals, use this radioactive carbon and when we eat crops or livestock the carbon becomes part of our cells and gets incorporated into our DNA where it stays. We know how much Carbon-14 was in the air before nuclear testing, and how it was eliminated after the Nuclear Test Ban Treaty outlawed above ground testing in 1963. It is therefore possible to estimate the turnover of cells. If someone born before nuclear testing shows no Carbon-14 in neuronal cells that suggests that no cells were added after birth. On the other hand if someone born at the peak of testing shows little or no Carbon-14 it would suggest that most of his or her cells had been replaced. Using this technique it has been found that cells lining the surface of the gut only last a few days and are among the shortest-lived in the body. Epidermal skin cells are recycled every two weeks or so. Red blood cells last about four months while liver cells turnover in six months to a year. The cells in the skeleton renew at the rate of about 10 per cent per year. Some cells last a lifetime. These are those in the lens of the eye, the oocytes (eggs) and most brain cells in the cerebral cortex.

5 Growth

"To live is the rarest thing in the world. Most people exist, that is all."
Oscar Wilde (*The Soul of Man Under Socialism*, essay, 1891)

The human foetus stays in the warm comfort of the womb for more or less 40 weeks. During this time it is bathed and cushioned by amniotic fluid at a comfortable 37°C. Oxygen and nutrition are provided through the umbilical cord that also disposes of waste material. The growing baby is of its mother, but is not its mother. It's a foreign body, a growing tumour within its host. The mother's immune system is designed to reject and destroy matter that does not belong. Mechanisms have evolved to ensure that the foetus is not rejected. When this mechanism fails, or is overwhelmed, problems can occur. Without drugs to suppress the immune system a body will reject a transplanted kidney or other organ, unless that organ is entirely compatible with its host. In the same way a mother can sometimes reject or harm the baby within her.

As the foetus matures it becomes recognisable as a human baby, and at some stage it becomes capable of living independently outside of its host. It becomes viable. Being born triggers a remarkable sequence of physiological and anatomical events. Lungs that have never taken a breath must expand to suck in air and then contract to expel carbon dioxide. The heart must reconfigure the channels through which blood flows to pump blood through the newborn lungs. From its cosseted hideaway the baby is suddenly expelled into a world of noise and suffering, confusion and pain. Everyone who has ever lived has gone through this process and all who ever have lived have also been destined to die. These days prenatal tests, ultrasound examinations and heart rate monitoring can usually tell the parents when abnormalities and problems are likely. But nonetheless every parent holds their new-born with wonder, checks the number of fingers and toes on each limb and counts their blessings that they have a normal child.

Of course, surviving outside of the womb does not make the baby capable of independent existence. Without someone to feed them and protect them a newborn would rapidly die. Indeed the child remains dependent on the parents for many years and would not survive without someone to care for them. It takes time for the child to be independent. Some skills are rapidly acquired, such as the ability to use a smart phone or recognise ice cream and chocolate. Other skills, such as helping with the washing-up or tidying a bedroom, may never be mastered. Nonetheless at some stage the ambitions of parents are usually achieved and the child leaves home. All too often these days they seem to bounce back, perhaps with children and without a spouse. Towards the end of life dependency often increases again.

The trajectory of ageing regularly results in those who are aged and infirm entering a care home of some sort or needing regular support to continue living at home. By this time they may well have lost the ability to be independent. Care homes are stuffed with residents incapable of looking after themselves, with age having robbed them of the skills to even dress, wash or eat.

Shakespeare in *As you Like It* writes about the seven ages of man from *"puking infant"*, through to *"old man, toothless, blind and helpless as a baby"*. We recognise the individual all the way from newborn to deathbed. With each individual there is continuity from the foetus in the womb, through to the babe-in-arms, the child growing up, the adolescent changing into the adult, and then the slow decline into old age accompanied by loss of physical and mental abilities. This individual or entity we call you or me, is linked by physical continuity and shared experience. Over time physical features, ability, attitudes and many other attributes will change. The adult is undoubtedly enormously different from the child but they are the same person stretched over the dimension of time. They look completely different but share the same genome. Although almost every cell in the old man is different from those in the baby, they have a common thread of physical being and experience that connects them from the first to the last.

As humans grow older they are constantly renewed. At my age there is very little left of the original me, the one that my mother gave birth to back in 1952. Although almost all of the cells of my body may have been replaced, I have existed in continuity from birth up until the present day. My body is the repository of my memories and experiences. They are unique. No one else shares them. On my body are the signs of ageing. Disease and trauma are written into my tissues with scars and furrows. But throughout it all, my individual genetic blueprint survives in the nucleus of almost all my cells. The cells differentiate and form a skeleton to provide structural support, muscles to enable movement and organs to nourish and sustain the body. Individual organs usually have several functions. Some of them are as follows:

The intestines stretch some seven metres from stomach to anus and carry out the job of absorbing the nutrients from what we eat and reabsorbing water to create stool. The liver processes digested food, neutralises toxins, stores iron and vitamins and synthesises proteins. The kidneys excrete waste and some drugs in the urine, manage fluid balance, help regulate blood pressure and control the production of red blood cells. The heart is a muscle pump that circulates blood throughout the body, and the lungs oxygenate the blood and remove carbon dioxide. The pancreas secretes enzymes to aid digestion and produces insulin and glucagon to control sugar metabolism. The spleen is one of the organs involved in the immune response and it also removes old red blood cells and holds a reserve of blood.

Several organs produce hormones. These include the thyroid and parathyroid, the adrenal and pituitary glands, the ovaries and the testes. Hormones are chemical messengers that circulate in the blood controlling body functions from hunger and emotions, to reproduction. Other chemicals called neurotransmitters conduct signals across synapses, the gap from one nerve to another. We are made from all these structures as well as a tongue, joints, ligaments, tendons, salivary glands, pharynx, oesophagus, stomach, gallbladder, diaphragm, ureters, bladder, organs of reproduction, blood vessels, lymphatic system, bone marrow and many other bits and bobs which I learned about at medical school.

The **brain** is most important organ of all. This is the repository of what makes us an individual. Graft a different brain onto my body and it will not be me. Take a genetically identical person, my identical twin or a clone, and he is not me. Take my limbs away and you will remove many of my functions but I will still be me. You can deprive me of my sight, my hearing, taste, smell and sense of touch. You can make me locked in, a brain within a skull on a body that is a life support machine. I will still be recognised as me. The inability to have an independent existence, or even be aware and interact with my surroundings does not prevent me being a recognisable individual. The babe in arms is me, the old person with severe Alzheimer's is me, the person in a persistent vegetative state is me, even if the bits of the brain that think, remember, communicate, feel emotion and desire are dead. Even though a fully functioning brain may not essential to being recognised as an individual, when we talk about prolonging life the central element is the brain.

If we want to extend life I would argue that it is the memories, personality, thoughts and emotions we would like to preserve, not a directionless hulk of a body that does nothing more than exist.

Until well into the 17th century most physicians thought that the ventricles, four fluid-filled cavities within the brain, were responsible for thought, sensation, emotion and understanding. It was not until the mid-1780s that the Italian physician Luigi Galvani demonstrated that electricity could activate nerves and muscles. Although in 1838 Theodor Schwann and Matthias Jacob Schleiden proposed that the cell was the basic functional unit of all living things, this was not generally held to apply to the central nervous system. It was only after microscopes and staining techniques had improved that the anatomy of the brain could be revealed.

Santiago Ramón y Cajal was awarded the Nobel Prize in Physiology in 1906 for his work on the anatomy of the brain. He showed that the brain is composed of discrete cells called **neurons**. The basic pattern for most neurons consists of three parts, the **cell body**, the **dendrites** and the **axon**. The dendrites receive impulses from sensory receptors or other neurons and send them to the cell body. The cell body contains the nucleus and sits between the dendrites and the axon. The axon is the long slender projection of the nerve cell that conducts electrical impulses away from the neuronal cell body to transmit information and instructions to muscles, glands or other neurons.

The adult human brain weighs 1.3 kg, about two per cent of body mass, but consumes about 20 per cent of total energy needs. Its substance can be broadly categorised into either grey matter or white matter. The grey matter consists mainly of the cell bodies of the neurons. Most axons are covered with a fatty substance called myelin. This provides an electrically insulating layer and is essential for fast conduction of nerve impulses. Loss of myelin is called demyelination and can cause diseases of nerve conduction such as multiple sclerosis and Guillain-Barré syndrome. The white matter consists of bundles of axons and gets its pinkish white colour from the blood-rich, myelin sheath.

The neurons are the electrically excitable cells that link to each other in an extraordinary and complex web of connections.

Crammed into the brain in a young adult there are said to be in total about 176,000km of myelinated nerve fibres. There are about 100 billion nerve cells each connected to around 10,000 others resulting in some 1,000 trillion connections.

Somehow the interplay of these connections creates consciousness, retains and recalls memories, controls the body and evokes emotions. It allows us to stare in wonder at the stars in the sky, to laugh with joy at the smile of a grandchild or to weep sad tears at the death of a loved one. Everything we know, or think we know, is filtered through our senses and only perceived or made sense of within the brain itself. The other cells in the brain are **glial** cells. By some measures there are more or less the same number of glia cells as neurons in the human brain. Others quote a figure of 10 times as many glial cells. They surround the neurons providing support and protection, and they also form myelin.

Humans self-designate themselves as the brainiest of animals. However the human brain is two to three times smaller than that of an elephant and four to six times smaller than that of some whales. We may, just, have the biggest cortex relative to the total brain mass but some other animals have a similar ratio. Some birds achieve primate-like levels of cognition even though their brains tend to be much smaller. The difference with our brains is that we have a large brain size compared with body mass, about six times greater than that predicted compared with other mammals.

We have some broad insights into the organisation of the brain. The outer layer is called the cortex and this is where thinking and movements are controlled. It is divided into two halves or hemispheres and each hemisphere is itself divided into four different segments or lobes. The frontal lobes (unsurprisingly to be found at the front of the brain) are largely responsible for judgement, problem solving and voluntary movement.

Behind the frontal lobe is the parietal lobe. This is where sensation, handwriting and body position are processed. At the very back are the occipital lobes that deal with sight. Underneath the other lobes are the temporal lobes whose job is dealing with hearing, language and speech. At the base of the brain, between the spinal cord and the rest of the brain, is the brain stem. Basic automatic functions such as breathing and sleep are located here.

The cerebellum is at the back at the bottom of the brain and is responsible for coordination and balance. There are further subdivisions within these structures and localised areas such as the thalamus and hypothalamus have specific functions. The lobes of the brain envelop four interconnected cavities called ventricles. The brain is covered with a membrane called the meninges and bathed in fluid called CSF, cerebrospinal fluid. But what is the brain's relevance to my study of longevity?

Perhaps it is not surprising that most brain cells last a lifetime. After all they are where the memories reside and control takes place of thinking, attitudes, emotions, moods and everything from political affiliation to religious beliefs. It used to be thought that neurons cannot be generated in the adult brain.

There is now some evidence suggesting that neurogenesis may continue in discrete brain regions in mammals, particularly in response to pathology such as following a stroke. Despite this, it appears likely that, in normal circumstances, almost all the cells responsible for higher cognitive activity are with us for life. This does not mean that the brain is unable to alter and adapt. Evidence suggests that brain structure and neuronal connections change in response to training and experience. For example, the volume of the hippocampus increases with practice of navigational skills, and motor and hearing areas increase in size in association with musical proficiency. You may want to learn how to juggle. Work at it and the grey matter in your brain responsible for voluntary muscle coordination will increase. Take a rest from juggling and these changes reverse. Intensive periods of learning can alter the medial temporal lobes and the posterior parietal cortex. Several studies have shown that the white as well as the grey matter respond to learning with anatomical changes.

Whether humans can live for hundreds of years without renewing the neurons in the brain is a fundamental question still to be answered? If we can find a way to replace neurons will we have wiped out memories, changed thinking and emotional responses and, in effect, created a different person?

More knowledge is accumulating about how brains work. Snails are not known for being the brainiest creatures but one study provided insights into how their decisions are made. Only two neuron types are needed to make choices about feeding. The first neuron reports the presence of food and the second decides if it is hungry. There is more and more understanding of how the brain receives information from the environment, and the underlying electrochemical processes that take place as it makes its decisions. For example, specific anatomical areas of the brain have been shown to respond differently in risk-adverse and risk-seeking individuals. The cortex in male youths with so-called conduct disorders is thinner than healthy controls and there are differences in brain surface area and folding. These changes may be related to empathy or social skills.

How much control do we, as individuals, have over how we behave and what we do if so much depends upon the pre-existing anatomy and chemistry of our brains?

Memory is an interesting phenomenon. Hyperthymesia is the rare, but well documented, condition where individuals can remember almost everything that happens to them. Some people have fantastic memories, often for particular events or facts. People with an eidetic memory can vividly recall images. We know there are effective training methods for improving memory. Memory is often partial and incomplete, and can be false with clear recollection of events that did not actually occur. Memory is fallible and is an active process and not necessarily that accurate. It's easily affected by distraction, alcohol and tiredness.

We have different types of memory. Typically, short-term memory can recall up to seven items for at most a minute. Unless we try to retain it this memory disappears. Long-term memory is where we store memories that we want to keep. Several types of long-term memory have been described. Explicit or declarative memory stores facts and events. Implicit or procedural memory is the unconscious memory of how to do things, such as riding a bike, playing the piano or tying a shoelace. We now know that different parts of the brain are responsible for different types of memory. Facts are encoded within the medial temporal lobes whereas procedural skills need brain areas required for movement. There are specialised areas within the frontal lobes for planning and organising behaviour, and the hippocampus contains neurons that encode location. The hippocampus is the site of short-term memory and the cortex for long term.

A change must be taking place in the brain for memories to be stored. Eric Kandel was awarded the Nobel Prize in 2000 for his work on the molecular biology of memory storage. A key part of the process is alterations happening in the synapses, the gaps between neurons. When brief and intense synaptic activity is followed by a prolonged increase in the strength of the synaptic connection this is known as long-term potentiation (LTP). A brief episode resulting in a prolonged effect is thought to be the basis of memory retention. And if the mechanisms are understood it is only a small step to be able to implant, delete or falsify memories.

The brain is the most complex and least understood organ in the body. Unlike the kidney whose functions are replicated with dialysis, an artificial organ cannot replace it. Unlike the liver, heart or lung, if it fails it cannot be substituted with a transplanted organ. If damaged it has little capacity for repair and regeneration. It is the most vulnerable of organs and will die if deprived of oxygen for more than a few minutes. About 15 per cent of people over the age of 80 have dementia and this figure may be increasing. At the moment there is no cure for dementia and only a poor understanding of its causes.

The brain and its functions must be preserved if individuals are going to live for hundreds of years and this may well be the major challenge and limiting factor in trying to achieve this goal.

6 Defining death

"I'm not afraid of death: I just don't want to be there when it happens."
Woody Allen (*Without Feathers,* 1975)

The human body is a complex, dynamic organism. If you're reading this you know what one is. You might think that it's easy to define death. Once upon a time it may have been straightforward. But peering into a possible future of extremely long-lived people it may not be as easy as you think.

The body consists of many specialised structures and organs that work together to ensure the body's survival. Some of the organs are essential to existence, others, such as the kidney, come with a spare in case of failure, and still other organs are dispensable when it comes to long-term survival. You will probably know that you can do without your appendix and tonsils, but have you ever wondered what your spleen is for, or why it can be totally removed? You may not want to lose your uterus, ovaries, testicles or prostate, but their loss is not usually life-threatening.

Dying is a process that usually begins when the heart stops and blood no longer circulates. Not so long ago death was easy to diagnose. If the heart or lungs stopped working, blood could not be oxygenated and circulated to vital organs and death swiftly followed. The diagnosis could be made by the absence of a pulse and cessation of breathing. One or other of these events was usually the end result of other causes of death as well. If both kidneys stopped producing urine, the liver completely failed to function, the pancreas stopped producing insulin or too much blood was lost, death would inevitably follow.

If we pause to consider the body as a collection of cells rather than a single coherent organism then it is true to say that all of the body doesn't die immediately. The brain is particularly vulnerable to being deprived of oxygen. More than three or four minutes of cerebral anoxia at normal body temperature will lead to permanent irreversible brain damage, coma and then death. If the main thinking part of the brain (the cortex) dies the body may continue to survive with basic functions such as breathing being controlled by the brain stem. The individual will not be conscious or capable of any thought or controllable action. This situation is called persistent vegetative state (PVS) and can only be reliably diagnosed if the patient remains like this for many months and any prospect of recovery has been ruled out.

This is different from brain stem death. The brain stem is the lower part of the brain that connects the spinal cord to the main parts of the brain. The brain stem controls breathing and other automatic reflexes. In the UK formally establishing that brain stem death has occurred is accepted as a diagnosis of death. By definition these patients must be on a ventilator (breathing machine). Their hearts beat, blood circulates and vital organs continue to work. These organs are therefore suitable for transplantation and are the main source of donors. Turn off the ventilator and, deprived of oxygen, these patients' hearts will soon fail. For death to be permanent it must be irreversible. In the past if the heart or breathing stopped that was the case. What has happened to change the situation?

These days intervention can often prevent death caused by sudden heart or breathing failure. Basic life support with chest compressions and mouth-to-mouth resuscitation is widely taught and seen regularly on medical TV programmes. As long as some oxygen can be supplied to the brain, medical intervention can often support and eventually restore both circulation and respiration. Getting oxygen into the lungs is important. Even more important is to make sure that the heart pumps so that oxygenated blood is delivered to the body. External chest compression squeezes the heart and results in a flow of blood around the body. In fact only in 1960 was this technique described in a scientific paper. Combining mouth-to-mouth breathing with external chest compression is the core of modern CPR (cardiopulmonary resuscitation). The final key component is the use of external defibrillation.

Abnormal cardiac rhythms are a common cause of cardiac arrest, and can occur even in patients who are young and healthy and have an otherwise normal heart. Defibrillation is applying an electric current to the chest wall of a patient to restore normal electrical cardiac activity. This technique also dates from the mid-1950s.

On March 17, 2012, Fabrice Muamba, while playing football for Bolton against Tottenham Hotspur, suffered a cardiac arrest. He received immediate life support from doctors who were at the game. It took 78 minutes before his heart was restarted after transfer to a specialist cardiac hospital. Although he has now retired from playing football his recovery has been amazing, and he has been able to take part in a Christmas edition of *Strictly Come Dancing* and go on to complete a degree in sports journalism. The key to his recovery was the short time from collapse until CPR was started. There were several doctors as well as paramedics at the scene so intervention started as soon as it was realised that something was wrong. Although early defibrillation could not restart his heart, oxygenating his lungs and providing external cardiac massage kept enough blood flowing through his brain so that, when his heart was eventually restarted, he did not have major permanent cerebral damage.

The heart is a muscular pump that circulates blood round the body. Blood permeates every tissue and brings life-sustaining oxygen and nutrients, and carries away waste products. During a biblical lifetime of three score years and 10 a human heart will beat about 2.5 billion times. During that lifetime the heart cannot take a rest, lie down and recuperate, or take time off. It must keep beating without pause or the body in which it sits will die.

The heart may stop beating because its electrical control mechanism has failed. As we have seen life support can keep the patient alive for a limited time until advanced medical intervention is available. If a defibrillator is to hand the heart can often be restarted with an electric shock delivered through the chest wall. One long-term solution is to implant a Pacemaker. These life-saving devices have been around for some 60 years and well over half a million have been fitted to British patients.

The blood supply to the heart itself may be compromised through narrowing of the arteries. This may be rectified with a bypass operation or angioplasty. If heart valves are damaged they can be repaired or replaced. Most patients having major heart surgery will be placed on a cardiopulmonary bypass machine. This takes the blood out of a large vein, through a pump and oxygenator and back into a major artery completely bypassing the heart and lung. Similar technology is found in artificial hearts that have been used with mixed success in a limited number of patients. When all else fails heart transplantation is a well-established procedure with good outcomes.

If a patient stops breathing in a hospital environment it should be no more than a routine emergency. In fact, it is very common to choose to paralyse patients during major surgery. If someone has stopped breathing because of a sudden onset of muscle weakness, mouth-to-mouth breathing, as described above, will get enough air into their lungs to keep them alive as long as the person doing the resuscitation can keep going. Any trained doctor, nurse or paramedic should be able to put a tube into a patient's airway and puff air into their lungs. Attach a mechanical ventilator to the tube and the patient can be ventilated indefinitely. This life-saving technique was developed for polio victims in Copenhagen in 1952. It works well for anyone who can't breathe because of muscle weakness including patients with high spinal cord injury and those, such as the late Stephen Hawking, with motor neurone disease. It is also useful for most patients with lung disease, such as pneumonia, or bronchitis. Today there are many people being kept alive in hospital and also in the community by artificial ventilation.

7 Organg failure

"My brain? It's my second favorite organ!"
Woody Allen (*Sleeper*, 1973)

Medical science has also come a long way treating patients whose other vital organs have failed and who would have died in previous generations. These days kidney failure is no longer a death sentence. Artificial kidneys will dialyse blood doing the same job as the kidney. Your kidneys perform a fairly simple job. Blood going to them contains waste products of metabolism including urea and uric acid. These are filtered out in nephrons within the kidney. The clever part of the kidney is to reabsorb nutrients such as glucose and amino acids, as well as the electrolytes sodium, bicarbonate, magnesium and calcium. Most importantly the kidney reabsorbs filtered water. About 180 litres of renal filtrate are produced each day but after water is reabsorbed this results in only about two litres of urine.

The common form of dialysis involves taking blood out of a large blood vessel and passing it through an external filtering machine before returning the treated blood to the body. This is called haemodialysis. Although the idea is fairly simple, these machines and membranes are now very sophisticated and the techniques have been refined over many years to ensure that the processes are now reliable and relatively risk-free. The kidneys have other roles including the release of hormones to control blood pressure, managing the production of red blood cells and vitamin D metabolism.

Kidney transplants are routine and common and, for most people with renal failure, a better option than dialysis. Many other vital organs can now be supported or replaced. Blood transfusions are, in effect, a transplant from a donor. Transfusion is routine and widely accepted. Each year in the United Kingdom more than two million units of blood components are transfused with very few adverse effects. Organ transplant is only possible because some organs such as the kidneys, liver and heart, can survive and recover function after a period deprived of a circulation. So called beating heart organ donations are from patients in whom brain stem death has been diagnosed. The donated organ, liver, kidney, lung etc., is being perfused with oxygenated blood up until the moment it is removed from the donor. However some organs are retrieved from patients whose heart has already stopped beating. In the period between removal from the donor and implantation in the recipient the organ is cooled, bathed in a preservative, and sometimes perfused through its blood vessels with a solution to prolong the time it can be kept outside the body. Even with steps to cool and preserve the organs, the time to transplantation is critical and is performed within a few hours.

Modern medicine is extremely effective in other conditions where the problem is well understood and a cure or replacement therapy exists. For example, without the manufactured hormone insulin, many diabetics who are living today would otherwise be dead.

As we have seen the brain is the one organ that is relatively poorly understood. Brain cells are metabolically very active and thus more vulnerable to oxygen deprivation than other cells in the body. There is not much now that can be done about repair or restoration of cerebral function. Replacement is out of the question. The brain is where all external inputs from sensory organs are received and processed. It is the also the organ that makes *H. sapiens* distinct from other animals, if only to the degree with which this organ is able to analyse, understand and manipulate our environment.

It is the organ that gives us insight into our own biology, and the organ that will provide the necessary information to extend our lifetimes beyond the 100 or so years that may be achieved today. The brain develops and also fails. A newborn indubitably has a functioning brain but it is many years before they have the capacity to understand the world around them well enough to be independent of their protectors. As we get old, brain function begins to decline, in the same way that many other organs deteriorate. We forget things, intellectual tasks get more difficult, memories fade and with age a high percentage of people will get severe problems such as dementia, so that the person that used to inhabit the body has faded and diminished until there is almost nothing of their essence and vitality left. Nursing homes and dementia units are repositories for these shells of human beings. They deserve our respect and support while they continue to exist.

In the right conditions brain cells can survive extensive deprivation. You may have heard of the Swedish doctor, Anna Bågenholm. In 1999 aged 29, she was skiing when she fell headfirst onto ice covering a frozen stream. She went through the ice and became trapped so that only her feet and skis were above the ice. Bågenholm remained conscious in the cold water for about 40 minutes before her heart stopped. By the time she was pulled from the stream she had been under the ice for 80 minutes. At that time her heart was not beating, she was not breathing and her pupils were widely dilated and not responding to light, normally a sign of catastrophic brain injury. Doctors on the scene started chest compression and mouth-to-mouth breathing. When a rescue helicopter arrived resuscitation continued and she was given oxygen. Defibrillation to restart her heart didn't work. She was extremely cold when she arrived at hospital. Normal body temperature is 37°C. Hers was 13.7°C. There were absolutely no signs of life.

Every doctor is taught that a patient cannot be pronounced dead if they are cold. They have to be warm before death can be diagnosed. Anna was connected to a cardiopulmonary bypass machine to circulate and oxygenate her blood. Over the next nine hours her body temperature was slowly increased to near normal. Her heart restarted but her lungs had deteriorated. She woke up 10 days later, initially paralysed from the neck down. She was artificially ventilated for 35 days and needed a further two months in an intensive care unit. She returned to work 140 days after the accident with no evidence of permanent brain damage. She is a testament to the potential for profound hypothermia to protect the brain.

Hypothermia is used to protect the brain in some medical situations. It is now common, although somewhat controversial, to cool a patient after they have suffered a cardiac arrest even after circulation has been restored. This is in an attempt to minimize any brain damage from the period when the blood supply was interrupted.

Profound cooling is also carried out during heart operations where, for technical reasons, the heart must be stopped and it is difficult or undesirable to use a heart lung bypass machine. Although the technique is used in adults the patients who get this treatment most commonly are very small babies. In these tiny patients the circulation and blood supply to the brain may be stopped for 30 to 40 minutes, or even longer, while the surgery is carried out.

We know that profound hypothermia can protect the brain for hours. Some people believe that protection can last for years. There are several companies who will take your diseased body after death and put it in a deep freeze in the hope that medical advances will be able to cure you in the future. After death the body begins to break down and lose its integrity as a whole organism. Some cells are more resistant to lack of oxygen than others. Reports suggest that skin cells may survive for up to 24 hours after death and sperm cells can maintain motility for 36 hours. There is a common misconception that hair and fingernails continue to grow. Dehydration causes skin to shrink and may provide the illusion that growth has taken place.

Cancer is caused by cells that grow out of control. In 1951, a woman called Henrietta Lacks died in Johns Hopkins Hospital in Baltimore, Maryland. Before she died samples were taken of her cervical cancer. These cancer cells were placed in a culture medium where they continued to divide and grow. The cell line produced from this is known as HeLa and continues to be grown in laboratories around the world. These cells are effectively immortal and continue to grow given the right conditions. This provides some evidence that death is not inevitable. Examination of these cells also provides insights into why some cells and organisms are longer lived than others.

What happens to all those microorganisms with which our bodies are populated and upon which we depend for our health and wellbeing? For many the death of the host is an opportunity and not a tragedy. As our body defences break down, released enzyme start to digest cell membranes, and bacteria and other invaders spread throughout our tissues. Our tissues are broken into component parts, back to the molecules and elements of which they were formed. They may then be taken up and incorporated into plants and trees, and eaten by animals.

In this way even though we die we may gain immortality, reincarnated to live again as a piece of another living organism.

8 Life expectancy

"The days of our years are threescore years and ten; and if by reason of strength they be fourscore years, yet is their strength labour and sorrow; for it is soon cut off, and we fly away."
The Bible, King James Version, Psalm 90:10

In the Book of Genesis, God makes mankind in his image and gives it dominion over other creatures. Part of the deal was giving men pretty long lives. Adam managed 930 years before he died and his descendants also lived for hundreds of years. The oldest of them was Methuselah who died at 969 years. Sadly the human race turned to wickedness and in Genesis 6 God has had enough and says "My spirit will not contend with humans forever for they are mortal; their days will be a hundred and twenty years." And to this day 120 years is more or less the very maximum that someone can live to.

Living to 70 or 80 years of age is commonplace though, which is in line with Psalm 90:10 where the Bible concedes a natural life span of three score years and 10, or fourscore if you're lucky. This may come as a surprise because a life span of 70 to 80 isn't so different from what we expect nowadays. It's not that long ago that average life expectancy for many people was around 30 years, or even less. Until fairly recently death was capricious, indiscriminate and ever-present. But average figures are very misleading. Historically death rates have been high for the young. Survive into adulthood and, baring accidents and injury, there was a good chance of living into old age. There is high mortality in the first few years falling away until it rises again as old age is experienced.

Thomas Hobbes writing in 1651 in his book, *Leviathan*, described a bleak world where the *". . . life of man, (is) solitary, poor, nasty, brutish and short"*. It may have rung true back then and for some it may still reflect their individual lives and circumstances. However, for the vast majority of people living today things are incomparably better. Back when he published his book average life expectancy in the United Kingdom was 39; it had changed little since the days of ancient Greece and classical Rome when it was about 35 years. Not much happened to alter those figures until the latter part of the 19th century. There was also huge inequality between countries and social classes in life expectancy. Unsurprisingly, on the whole, the wealthy fared better than the poor and uneducated. In 1840s Whitechapel, in East London, the upper classes lived for an average of 45 years, tradesmen for 27 years and labourers and servants for only 22 years.

By the beginning of the 20th century average life expectancy was starting to increase in the richer industrialised countries, although in India, South Korea and other poor countries an average citizen could expect only to live for 20 to 30 years. During the last 100 years average worldwide life expectancy has more than doubled with life expectancy in South Korea now being greater than that in the UK. Most of this reduction in mortality has occurred since 1900 and has thus been experienced by only the last four generations of the roughly 8,000 human generations that have ever lived.

One of the most extraordinary things is that this benefit has affected all age groups, all social classes, and every country in the world. For the first time in human history the benefits that confer a long and healthy life have spread out from the elite to the masses. Even those living today in the countries with the lowest life expectancies have a longer average lifespan than those living in 1800 in countries with highest life expectancies. Average life expectancy is now more than 80 years in many countries including the United Kingdom.

Being born is still risky, although today much less so than in the past. Figures from developed countries show that the number of stillbirths, or late foetal deaths, ran at about three per cent to four per cent from the early 1800s right up until the 1940s. The figures are now less than 0.5 per cent, or about one in every 200 births. Every stillbirth is a personal tragedy and hopefully more will be done to lower this figure yet further. Medical advances in the care of neonates have dramatically improved the survival chances of premature babies. The duration of a normal human pregnancy is more or less 40 weeks, at which time the baby is said to be at term.

In 1800 more than 15 per cent of British children died before their first birthday and one-third died before their fifth. By 1900 although the figure for deaths within one year of birth was little improved, about 80 per cent of children survived to reach the age of five. In the first half of the 20th century there was a dramatic improvement in medical care, particularly in the treatment of infectious diseases. In 1950 four per cent of children died before they were five and that figure has continued to fall so that today it is fewer than one in 100. It is not just the rich countries that are succeeding.

In 1960, eight years after I came into the world, being born was still a chancy business in many countries. Global child mortality was over 18 per cent. Despite the naysayers, in many ways my generation has lived through a golden period of world history; we are living through a unique period of time. There is no precedent for the lives we are living and the changes taking place around us. Although there is still much to do, current global child mortality has fallen to about four per cent, an astonishing and concrete example of what can be achieved. There are millions of children growing to be adults who would have died not so long ago. As recently as the year 2000 it is estimated that 9.9 million children worldwide died before they were five. By 2013 the number had fallen to 6.3 million, still far too many, despite the number of births increasing by some 10 million. Rising prosperity, better nutrition, sanitation, education and health care are the reasons for this improvement.

Nowadays, newborns dying within the first month of delivery often die from prematurity, congenital problems or complications of delivery. After that the main causes are still infection. But in bygone times infection would destroy whole communities in a matter of days. The big killers were tuberculosis, scarlet fever, smallpox, influenza, typhoid, cholera and sexually transmitted diseases. Thankfully most of these deadly killers are rarely a cause for concern these days. The bubonic plague that hit Europe between 541 and 542AD wiped out around 40 per cent of the population. The second pandemic of the late 1340s was known as the "Black Death" and killed 30-60 per cent of the population of Europe causing some 100 million deaths, or round about a quarter of the total population of the world at that time. These pandemics are not confined to ancient historical times and the flu outbreak of 1918 to 1919 killed an estimated 75 million worldwide. Infection killed *en masse* but was also an ever-present killer on a more down-to-earth and personal level. Before the era of antisepsis, immunisation and antibiotics, death from infection was a constant threat that haunted all families from high to low.

Currently about 200 people per year die in workplace and industrial accidents in Britain. During the industrial revolution unsafe machines, explosions or tunnel collapse killed thousands every year, including children. Deaths on the roads disproportionately affect the young although the numbers are tiny compared with the millions carried away by preventable disease. In the United Kingdom, road deaths peaked in 1966 with nearly 8,000 deaths a year. In 2015, despite a huge increase in the number of vehicles on the road, deaths had fallen to 1,700 annually. Road traffic accidents are still the leading cause of death in five to 19-year-olds with suicide and homicide being the second and third most common reason. Suicide remains the commonest cause of death in those over 19 until heart disease and cancer take over in those older than 50. Young people bear the brunt of deaths in wars. The estimated number of deaths worldwide in World War I was 20 million, and in World War II, getting on for 60 million. Pollution from burning coal was a widespread cause of respiratory problems, as was the high incidence of smoking tobacco. Both of these risks have diminished enormously during my lifetime. To replace them we are now much more likely to have to contend with heart problems, cancer, dementia, obesity, diabetes, hypertension and degenerative diseases.

The increase in life expectancy has not only been because of the fall in child mortality. In the UK in 1845 a five-year-old could expect to live a further 50 years. Today's five-year-old can expect 27 years more than that with a life expectancy of 82 years. These benefits continue at yet higher ages. A 50-year-old of 150 years ago might expect an additional 20 years of life. Now the figure is more like 35 years. These improvements are thought likely to continue. The predicted life expectancy of a child born today is some five years longer than that of a child born in the early 1990s. If progress at reducing mortality continues for those who live in countries that already have high life expectancies, then most children born since the year 2000 will live to celebrate their 100th birthday.

The effects of 150 years of social engineering, wealth creation and medicine are that almost everyone lives longer and will continue to do so.

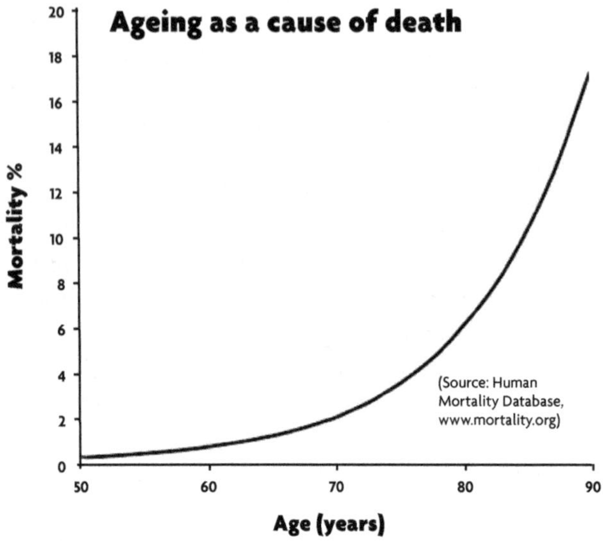

Ageing as a cause of death
20
18
16
14
12
10
8
6
4
2
0
Mortality %
50
60
70
80
90
Age (years)
(Source: Human Mortality Database, www.mortality.org)

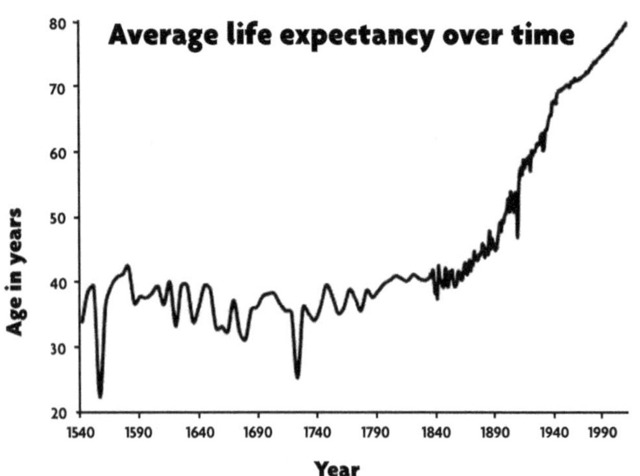

Average life expectancy over time
80
70
60
50
40
30
20
Age in years
1540
1590
1640
1690
1740
1790
1840
1890
1940
1990
Year

9 Historical longevity

"Death is only the end if you assume the story is about you."
Night Vale podcast (www.welcometonightvale.com)

Humans are very long-lived compared with most animals. Do the figures reflect the fact that many more people are reaching their possible full lifespan or are people actually living longer than was achievable in previous times? There is some interesting information from the far distant past. Although clearly not typical citizens, those Kings of Judah between 1,000 and 6,000BC who did not die a violent death (and there were a few of those) lived for an average of 52 years. Although King Saul was killed in battle his son, David, reigned for 40 years and died a natural death aged about 70 to 75. David's son Solomon also died of old age aged about 80. His son Rehoboam died a bit younger at 58 years, however, his 18 wives, 60 concubines and many children may have sent him to an early grave. Greek philosophers, poets and politicians living between 450BC and 150BC managed an average of 68 years. Socrates lived to the age of 70 before dying in 399BC. Mind you if he hadn't allegedly drunk a mixture containing hemlock he may have gone on for a few more years. Plato died of natural causes in 347 or 348BC at about the age of 80 and Archimedes was about 75 when he died in 212BC, apparently not in his bath. Pythagoras was also about 75 when he died (for real, not theoretically) in 495BC.

Long life is not that unusual. Galileo Galilei was born in 1564 and lived for 77 years. Isaac Newton was 84 when he died, and Michelangelo made it to 88. What of Zeus, King of the Gods? Well, he's immortal so he's still around. The ancient Romans did slightly less well but some have attributed that to the use of lead piping causing chronic poisoning. Cicero made it to the age of 63 before being assassinated in 43BC and Plutarch was 74 when he died in 12AD.

Further evidence comes from examining the lives of ancient Greek and Roman men whose dates of birth and death are recorded in the Oxford Classical Dictionary (2nd edition). These men had to have survived childhood and been notable, and they were probably wealthier and better fed than much of the population. Even so a quarter met a violent end by assassination, forced suicide or death in battle. The median length of life of those who died of natural causes was 70. So looking at this, admittedly highly-selected bunch of male ancient citizens, if you got through childbirth, reached adulthood and didn't die a gruesome death you had a good chance of getting to 70. In fact, you had a 50:50 chance of living longer. Of course these figures apply to men. Since childbirth ceased being such a hazardous event, women have consistently managed to outlive men.

In England, examination of skeletal remains show that people got taller during Roman times and then shrunk in the Dark Ages of the early medieval period. They then increased in height after 1000AD before shrinking in the centuries after the year 1200 covering the periods of the Great Famine in Europe (1315-1317) and the Black Death (1346-1353). People then appeared to get taller again before becoming shorter once again on the eve of industrialisation. Average heights between 1400 and 1700 were similar to those of the 20th century, while those in the early 19th century were shorter than Roman times.

If average heights reveal something about health and nutrition then perhaps it's not surprising that some people in the ancient world were able to reach what we would now consider to be a ripe old age. Going on current figures a man or woman can reasonable expect to live well into their eighties, and this is clearly an age which has been attainable from ancient times. The big difference today is that many more people are getting to this age.

In Britain, the Queen used to send a telegram to any of her subject reaching the extraordinary old age of 100, a tradition that started with King George V. Nowadays a card is sent rather than a telegram, and perhaps in the future it will be an email, text or posting on social media. With increasing longevity another message will be sent on the 105th birthday and each subsequent anniversary.

Although extremely uncommon, I believe it likely that a few people made it to live to one hundred years old in previous centuries. It has been claimed that in January 1831 there were 16 centenarians in Belgium, after all many of the factors we today associate with a long life existed well before the modern era. If food was adequate and nutritious, and disease could be avoided, a non-smoking, outdoors, active lifestyle is thought to be beneficial. In 1981 there were 2,600 centenarians in the UK. In 2015 this figure had grown to 14,750 and by 2050 the number is projected to be more than a quarter of a million. In the United States of a cohort of 100,000 men and women born in 1900, 0.5 per cent of men and 3.2 per cent of women lived to see in the millennium in the year 2000. Predictions for those born in the year 2000 are that at least 15 per cent of men and 20 per cent of women will reach 90 years of age. As more people are reaching these very old ages it appears that deteriorating health is being delayed, not merely stretched out over more years.

Worldwide there are thought to be about 400 people who are older than 110 years, supercentenarians, and more than 90 per cent of them are women. To date the oldest verified supercentenarian is Jeanne Calment, a French woman who died in 1997 at the extraordinary age of 122 years and 164 days and, so far, she remains the only person who has lived beyond 119 years of age.

Not only are the supercentenarians very old but, given their extreme age, they tend to be pretty spry. In one study from the USA 40 per cent managed well with minimal or no assistance and fewer than 15 per cent had a history of vascular-related disease such as narrowing of the arteries to the heart. In Japan 83 per cent did not have problems with any major diseases until they were 105 or older. In contrast, mere centenarians tend to get medical problems such as heart disease, stroke, hypertension and dementia, in their nineties, which is still pretty good going but 10 years or more earlier than those who will pass the 110 mark.

Looking at major diseases such as cancer, dementia and stroke, as well as the decline in mental function, the supercentenarians develop these problems later than the centenarians who, in turn, stay healthy longer than those who only make it into their nineties. Indeed, in one study some diseases, including Parkinson's and diabetes, are very rare among the most elderly. Brain atrophy, loss of cortical thickness, is common with ageing. However, there are some older adults who, based on memory tests, maintain brain function as well as younger people. Compared with their peers they show little or no loss of cortical brain volume in the regions that are important to memory storage, attention, encoding and retrieval. Yet more evidence that the extremely old are different from the rest of us.

With present knowledge it seems likely that living into your 80s or 90s is within the reach of most of the population. The current limits of human longevity are around about 120 years of age. What is less clear is whether the supercentenarians live for so long because they have a genome bestowing favourable qualities, or because they have a life style and live in an environment that would give the rest of us a good shot at reaching this age.

10 Ageing: Visible effects

"Eternity is a terrible thought. I mean, where's it going to end?"
Tom Stoppard (*Rosencrantz and Guildenstern are dead*, Act II, 1967)

Most living organisms age. In many ways ageing can be thought of as a metabolic clock programmed to count down the years of our lives. It is the effect that the passage of time has on us, and our bodies. Despite very different life expectancies a two-year-old mouse, a 12-year-old dog, a 30-year-old chimpanzee and an 80-year-old man show many common signs of ageing. They share deficits such as loss of tissue elasticity and muscle strength, impaired sensory perception and reflexes, memory loss and age-associated disease. What is very different is the time scale over which these changes develop.

After we are born we develop, gain knowledge and experience, pass through puberty with its dramatic physiological changes, and reach adulthood. Many of the processes associated with ageing are known and understood. However, there is, as yet, relatively little consensus as to the direct causes of ageing. Most proposed treatments are therefore aimed at the changes seen in the aged rather than a targeted intervention directed at the reasons why we age.

The essential lifespan of humans is about 40 to 50 years. This gives time to reproduce and bring any offspring to maturity. In the wild, most animals die from predators, disease, starvation or drought long before old age sets in. The ageing process was only revealed once ways of extending life expectancy were discovered. The increasing number of older people can be seen as an artefact of human civilisation.

Survival is a dynamic balance between the occurrence of age-related damage and properly functioning maintenance and repair systems.

Evolutionary natural selection is geared towards attaining the essential lifespan. Once this is achieved there is less of a need to maintain a healthy organism. Huntington's Disease is an example of a dominant lethal mutation that persists because the late onset of the disease (30 to 40 years) allows a carrier to reproduce before dying.

The disease therefore escapes the pressures of natural selection. In the same way an accumulation of late mutations that do not exert an evolutionary pressure may contribute towards ageing. There is some support for the theory that genes are selected for beneficial effects early in life although they may have unselected harmful effects with increasing age.

There is undoubtedly also a genetic component as to when and how we age. Although single point mutations can cause syndromes with the features of ageing, the genetic contribution to human ageing is thought to be diffuse and ill understood, and may be partly determined by accidental and random effects. Ageing can probably best be seen as a complex multifactorial process. How much is programmed is open to debate. We know there are biological clocks that, to some extent, regulate the timetable of a lifespan through growth, development, maturity and old age. This depends on genes switching on and off altering endocrine and immune systems. In mice, hypothalamic stem cells have been shown to govern the speed of ageing. With increasing age the number of hypothalamic stem cells decline and this decline accelerates ageing. Amazingly, if healthy hypothalamic stem cells are implanted ageing is slowed and lifespan extended. The mechanism is through the expression of microRNAs that play a part in regulating gene expression.

Ageing may also have something to do with the ability of cells to divide and renew themselves. For some cells there is a limit to the number of times that they can divide. Telomere shortening may be relevant and is addressed in greater detail later. Undoubtedly, over time, there is a reduced ability of cells to cope with stress leading to an increasing risk of disease and death.

We live in an environment teaming with microorganisms and they may be part of the ageing process. As humans age the defences for keeping microbes out weaken. Junctions connecting cells loosen and the immune system tires. Older adults have higher circulating values of lipopolysaccharide (LPS) that is likely to be of bacterial origin. An injection of LPS causes metabolic effects typical of those seen with ageing. Viral infection has been shown to shorten telomeres and cause atherosclerosis. Compared with matched controls, patients with Alzheimer's have higher numbers of circulating bacteria, viruses and fungi. In an animal model the size and incidence of cancer is greater in the presence of LPS.

Several studies have been carried out on germ-free mice. The gut microbiota affects brain signalling and behaviour, and appears necessary for normal stress responses, sociability and cognition. Germ-free mice are not immortal but they have been shown to live 20 per cent to 50 per cent longer. Germs, just like oxidative stress, mutations, bumps and bruises, are another factor related to living that may take its toll on us.

There is a functional decline with age, some of which starts shortly after reaching maturity, while other abilities are maintained until later in life. While sexual desire can be maintained for some time, maximum fecundity – the ability to produce offspring – peaks in the mid to late twenties for men and women. This is a complex of various factors. It includes frequency of sex, ovulatory cycle and duration of fertile period as well as sperm concentration, motility, morphology and seminal volume. You may read that male fertility peaks in late teens. However, men can't produce any babies unless they get the opportunity to sow their seed.

Most deterioration begins, subtly at first, from the early twenties onwards. The outward appearances are how we estimate someone's age, something that can usually be done quite accurately. Ageing reveals itself in the wrinkles we develop, and the greying and loss of hair (on the head for men – it seems to envelop the ears and grow rampant in the eyebrows). Close objects become blurred because of hardening of the lens in the eye. Joints ache, bones thin and become brittle. Muscle mass decreases and strength and mobility decline. Hearing commonly deteriorates as does memory and mental agility. The risk of getting cancer, heart conditions and degenerative diseases increases.

By their mid-thirties, a Premier League football player is over the hill, unable to compete with youngsters, despite having the best possible nutrition and conditioning. Times for competitive swimmers are maintained until they reach their mid-30s and then progressively decline until the age of around 70, before dropping off precipitously.

Peak muscle strength is reached at around 30 and remains stable until 40 to 45, following which it starts to decline. Between the ages of 25 and 80 as much as 40 per cent of muscle mass can be lost, and functional loss of strength is often greater than the percentage of muscle loss. Cardiac output, the amount of blood the heart pumps round the body each minute, decreases by about 1 per cent a year after the age of 30 so that an 80-year-old has about half the output that they had in their prime.

Arteries thicken and harden and blood pressure tends to increase over time being a significant risk factor for stroke and heart disease. The elastic recoil of the lungs decreases and there is a tendency for the airways to collapse. Lung function progressively decreases at about one per cent per year from the age of 25 years onwards. This functional decline is accompanied by structural alterations including enlarged airspaces, a fall in surface area and a decrease in elastic recoil. Old people can't breathe as deeply and they are also more likely to catch chest infections. Kidneys shrink in size and also in the number of active filtering components. Their functional ability fades affecting the excretion of drugs such as digoxin and gentamicin, and the ability to reabsorb glucose and water. Urinary incontinence becomes common and, in men, the prostate enlarges.

From the oesophagus to the colon there is a loss in intestinal motility and sphincter control. The liver has a large reserve capacity. Although it typically shrinks with age it usually manages to cope with the normal tasks it is given. Obviously many years of abusing it will hasten the time when it can't handle the toxins that come its way. Also, over time the beta cells in the pancreas become impaired and the incidence of insulin resistance increases with age. Bones thin and become weak so that by the eighth and ninth decades some 30 per cent to 50 per cent of skeletal mass may be lost.

There are also endocrine and immunological changes. For women the menopause ensues, and for all of us age-related diseases such as dementia, atherosclerosis and arthritis, become more likely. Although symptoms are similar there are several recognized underlying pathologies resulting in dementia. The commonest is Alzheimer's accounting for more than 60 per cent of cases. Nearly 20 per cent are related to vascular problems while others are associated with other conditions including Parkinson's disease, depression, multiple sclerosis and Creutzfeld Jacob disease. Some diseases may become less of a problem. Asthma is one example, and sciatica is less likely as vertebral discs dry up and lose their jelly-like properties.

At some point in middle age, getting older is associated with an increased risk of becoming seriously ill. Death rates go up until eventually no one is left. Interestingly, chronological age is not accurately tied to physiological changes. Chronological age – the passage of time – is not the same as biological age.

Biological age has been estimated from markers based on physiological data such as cardiovascular, metabolic and immune function. By this measure those who age more rapidly show earlier cognitive decline, worse health and look older.

11 Ageing: Cellular effects

"Life is not a problem to be solved but a reality to be experienced."
Søren Kierkegaard (unknown source)

CELLULAR CHANGES

With ageing, changes are also seen at the cellular level. Some of these may be programmed into our bodies. Humans do not appear to be designed to live past a given number of years and the result is, as a species, that we do not outstay the time allotted to pass on our genes, nurture our young and then make way for our descendants. Over time our cells are subject to all sorts of insults (physiological ones that is) and stresses. These make their marks on our cells and us.

Old cells look different. They are usually bigger with measureable changes in enzymes such as ß-galactosidase. Gene expression may alter. Ageing is associated with damage to DNA, and also potentially harmful epigenetic changes through mechanisms such as histone methylation. There is attrition of telomere length. Errors in protein manufacture and processing become more common. There is exhaustion of stem cells and a breakdown in intercellular communication.

Mitochondrial dysfunction increases with age. Mitochondrial DNA is close to the sites of oxygen radical production and unprotected by the histones that are found in DNA within the nucleus.

Longevity has been correlated more closely with age at maternal death than of paternal death. This suggests that mitochondrial DNA inheritance may have an important role in determining how long someone lives.

Over time mitochondria become less efficient and there is more mitochondrial DNA damage and production of oxygen free radicals. Free radicals are unstable highly reactive atoms or molecules with an unpaired electron. There are changes in mitochondrial shape, volume and length, normally seen as an increase in size and a more rounded shape. Maximum energy production seen in the ability to produce ATP falls by some eight per cent per decade. In the elderly mitochondrial activity may be half that of young adults (ATP [adenosine triphosphate] is the high-energy molecule that stores the energy we need to do just about everything).

In human cells **superoxide dismutase** is a key antioxidant. An imbalance between oxygen radical production and antioxidant defences has been identified as an important factor in ageing. The damage caused by free radicals affects many structures including DNA, protein and lipid. Minimising the amount of oxidative damage may be important as demonstrated by, for example, the increase in the lifespan of flies by boosting superoxide dismutase and the fact that long-lived worms are relatively resistant to oxidative stress. The importance of oxidative damage is not yet firmly established as shown by the lack of effect on longevity of manipulating superoxide dismutases in the worm *Caenorhabditis elegans (C. elegans)*. In humans, commonly studied antioxidants include beta-carotene, vitamins A, C and E and selenium. However, a recent authoritative review failed to find any evidence that antioxidant supplementation had any benefits in humans. On the contrary beta-carotene and vitamin E were associated with an increased mortality, as were higher doses of vitamin A.

SENESCENCE

At the cellular level **senescence** is the mechanism that limits cell division to a finite number. Senescence is an important and beneficial part of wound healing, injury repair and defence against cancer. The number of senescent cells increases with age. This may be because more of them are being made or because the clearance mechanisms start to fail. If senescent cells are not efficiently cleared they linger and adversely affect cellular function and the ability to respond to stresses. This is because senescent cells secrete a potentially toxic mixture of chemicals and enzymes that seep out and affect nearby cells. These chemicals include "mammalian target of rapamycin" (mTOR), members of the interleukin family of cytokines, matrix metalloproteinases, insulin and insulin-like growth factor-1 (IGF-1) and various proteins which convert or transcribe DNA to RNA. Other relevant mediators include calcium and sirtuins (proteins which regulate the ageing and death of cells and their resistance to stress).

In a study in mice it has been shown that killing senescent cells increases median lifespans by about 25 per cent and also delays progression of cancer. In addition many age-dependent changes occur more slowly. No observable detrimental effects were seen. Because of their apparent involvement in many ageing and disease processes, eliminating senescent cells and attenuating the associated pro-inflammatory response is a target for some anti-ageing treatments. This may explain why vulnerability to diseases and the incidence of chronic conditions increase with age.

A minute worm has made an important contribution to understanding ageing. *C. elegans* has a fixed number of cells in its adult form. Hermaphrodites have precisely 959 cells. Each of the 959 cells develops in a fixed and predictable manner. Not only that, but during its growth precisely 131 extra cells are produced and then killed. This is programmed cell death; something called **apoptosis**, and is essential to the development of the adult worm. Apoptosis is a normal part of development and ageing. Initially cells shrink and become engulfed by macrophages – the dustcarts of the body that digest and clear away cellular debris. Unlike cells that swell and burst and die of necrosis (sometimes called oncosis), there is no leakage of the cell contents and thus no inflammation or harm to surrounding cells.

Necrosis is an uncontrolled, passive process that usually affects large numbers of cells. Apoptosis is energy-dependent, targeted and controlled by genes. The main ways of initiating apoptosis converge on the same final pathway where caspases – enzymes that break up proteins – are activated. Every day billions of cells are made just to balance out those dying by apoptosis, and that number is much greater during normal development, ageing or with illness.

Both the nervous system and immune system overproduce cells during development. This is followed by the apoptosis of those cells that fail to make working synaptic connections or effective antigens. Apoptosis is also the way that the body gets rid of cells that have been occupied by bacteria and other pathogens. It is an important part of the process of wound healing removing excessive fibrosis and scarring. Sometimes apoptosis can be suppressed and this is a feature of some cancers. Excessive apoptosis may occur in autoimmune diseases, neurodegenerative diseases and ischaemia-associated injury where there is insufficient blood supply to parts of the body, especially the heart muscles. Therapies are currently being developed to target overexpression.

Telomeres

The ends of chromosomes are protected by a cap of repeated non-coding DNA sequences and proteins called telomeres. The DNA sequence in humans and other vertebrates consists of TTAGGG. At conception the length of a telomere averages about 15,000 bases, some 2,500 TTAGGG repeat units. They protect the chromosome from recombination and from end-to-end fusion. They are necessary for complete replication of the chromosomes, help regulate gene expression and function as a molecular clock. When cell division takes place the telomere shortens by between 30 and 80 bases per cell division. At a certain length, about 5,000 bases, the cells lose their ability to divide.

Once the telomere has shortened to a critical length, the damage-repair mechanism sees the unprotected DNA double strand and starts a process leading to senescence or apoptosis. **This is an important process to prevent cells turning cancerous**. Senescent cells are no longer capable of replication and shut down their metabolism to a minimum. Rarely cells are able to maintain telomere length, usually through activation of the enzyme telomerase. This may lead to immortalisation when cells are apparently able to keep on proliferating.

Telomerase maintains telomere length by adding telomere repeats to the ends of chromosomes and is one of the mechanisms by which cancer cells can continue to grow unchecked. Most normal adult human cells do not contain any telomerase but it is present in most cancer cells.

Telomere length is highly heritable in humans. It is also influenced by ethnicity and gender. Longer telomeres are linked to a healthy lifestyle and physical activity, and shorter ones with environmental and psychological stress, smoking and obesity. Shorter telomeres

are also associated with hypertension, heart disease, diabetes and Alzheimer's. Telomere length positively correlates with healthy years of life and are longer in a group of centenarians compared with "control" men and women. Breed mice with short telomeres and premature deaths are seen. Restore telomere length with telomerase and degenerative changes are reversed.

These properties strongly suggest that telomerase activity and telomere length play important roles in cellular ageing and the development of cancer. Indeed cells that have been supplied with an external source of telomerase stay young and proliferate indefinitely. These cells are not only immortal but show ageing reversal.

Sadly the secret of eternal life is a bit more complicated than simply inducing telomerase. There is a relatively poor correlation between telomere length, chronological age and functional deterioration. There appear to be many other much better predictors of mortality. Although, in general, telomere length shortens with age, at least one study has found that it can remain stable or even increase in later life. Other mammals, including mice and rats, have short lives and long telomeres. Of all primates studied, humans have the shortest telomeres and the longest lifespans. Although mice have long telomeres they shorten 100 times faster than human ones, and the telomeres in dogs shorten 10 times faster.

These findings suggest that it may be the ability to preserve telomere length rather than length itself that may be important.

Stem Cells

Stem cells are the cells that we all start with back in the embryo. They contain all the genetic information to become any type of cell. During development, through activation of specific genes, the embryonic cell differentiates and turns into the specialised cell that is its destiny. Human embryonic stem cells can be derived by taking a single cell from an eight-cell stage embryo. The remaining embryo can still grow and divide normally. Stem cells have been used therapeutically for some time. A bone marrow transplant is, in effect, a form of stem cell transplantation. Chemotherapy destroys the ability of the recipient's bone marrow to produce new cells. A bone marrow transplant from a suitable donor will restore the ability of the bone marrow to produce new blood cells.

A pluripotent stem cell is a master cell with the ability to turn into almost any other cell type in the body. The discovery that mature cells can be turned back into pluripotent stem cells was important enough to be worthy of the Nobel Prize for Medicine, which was awarded in 2012 to Sir John Gurdon and Shinya Yamanaka. Their extraordinary discovery was that the differentiated adult cell still possesses the information to turn itself back to the embryonic state to form an induced pluripotent stem cell (iPSC). All it needs is for the right genes to be activated. This technology avoids the use and destruction of human eggs or embryos.

The usefulness of making pluripotent cells are obvious. As they can be generated from an individual they can be made in a patient-matched manner. These cells could replace those lost or damaged, and could be used to build new organs without the risk of rejection. Stem cells are being investigated to replace or renew damaged tissue in brain and spinal cord, in the heart, to replace teeth, restore vision and transplant new pancreatic cells to produce insulin. So far pluripotent stem cells have been found to be safe in humans and there is growing impetus to do further clinical testing. The major concern with the use of iPSCs is their tendency to form tumours. There are many other potential uses that are being looked at. As well as therapeutic use in patients, pluripotent stem cells have great advantages for creating *in vitro* models of various conditions. They are also being used to develop and screen drugs and to study the role of genes in disease.

Many other features have been noted in ageing cells, although it is often hard to distinguish between cause and effect, to separate the signs of ageing from what causes it. In summary, there are many overlapping mechanisms that may all contribute in various ways to what we see as ageing. If ageing is to be cured it is therefore likely that many different strands of this problems will have to be attacked. Like a particularly leaky plumbing system you may plug up one hole only to find that the water is escaping from another hole elsewhere in the piping.

Ageing is a reflection of cellular damage but also a failure of the body to repair and replace damaged cells. The rate of progress is tremendous. Many more studies are in the pipeline and the role of stem cell therapy and other interventions will become much clearer over the next few years.

12 Ageing: Genetics

"There's an old joke, um . . . two elderly women are at a Catskill mountain resort, and one of 'em says, 'Boy, the food at this place is really terrible.' The other says, ' Yeah, I know, and such small portions!' Well, that's essentially how I feel about life – full of loneliness, and misery, and suffering, and unhappiness, and it's all over much too quickly."
Woody Allen (*Annie Hall*, 2001)

*H*omo *sapiens* has been remarkably successful in spreading out over almost all the globe. As a species we have managed to thrive in an enormous range of habitats, adopted many different lifestyles, and been sustained by a wide variety of diets. Throughout the world a lifespan into the 80s and 90s is achievable with access to adequate nutrition, clean water, sanitation and medical support. Despite 28 years in prison with poor conditions and illness, Nelson Mandela died aged 95. Castro was 90 when he died in 2016 despite being involved in revolution and smoking cigars. Kirk Douglas celebrated his 101th birthday on December 9, 2017 demonstrating that a Hollywood lifestyle does not always lead to an early death. In the UK many more people are surviving until 100 and getting a card from the Queen.

My reading of the evidence suggests that the supercentenarians are different from the mere striplings who make it to their 80s or 90s. Their average age at death is well beyond two standard deviations from the norm. If this is true then most of us have the potential for a long and healthy life, but many fewer have the genes to make it beyond 100.

There is a clear genetic component linked to longevity with between 20-30 per cent of duration of lifespan attributed to genes, and the longer you live the more important the genetic component seems to be. People who live extremely long lives are less likely to have alleles (gene variants) that predispose to common diseases including heart problems, Alzheimer's, high cholesterol and chronic kidney disease.

Single nucleotide polymorphisms (SNPs) are the commonest type of genetic variation. Several SNPs have been found that are more common in the very old.

A gene called SERPINE1 encodes a protein called plasminogen activator inhibitor-1 (PAI-1) that regulates cellular senescence. Researchers found that some members of the Amish community living in Indiana, USA, have a rare mutation in this gene causing a lifelong reduction of about 50 per cent in the protein. This mutation is linked to lower fasting insulin levels, a lower prevalence of diabetes and longer white cell telomeres. The Amish remain relatively isolated from outside influences and the genetic abnormality can be traced back to a common ancestor from the late 1800s. The median lifespan of those with this gene is 10 years longer than unaffected individuals. This study provides powerful evidence that longevity is strongly influenced by small genetic differences. This is encouraging for investigators who believe that seeking a cure for old age is a realistic goal. For those without this mutation, drugs that selectively inhibit human PAI-1 may have a similar effect and are already being investigated in human subjects

On the whole if your parents lived a long and healthy life your chances of doing so are quite good. Siblings of centenarians are much more likely than the general population to reach 90 years or more. Those with long-lived parents age more successfully than others, maintaining objective and subjective physical function for longer. Close relatives of extremely long-lived people are twice as likely as others to also have a long life. The children of centenarians tend to be older before they get chronic age-related conditions such as heart disease, diabetes and stroke.

Gender is clearly a genetic condition. Allowing for the risk of childbirth, women on average, live longer than men. To live a long life you must also avoid a whole medical textbook full of inheritable diseases. About three per cent of human diseases are caused by a defect in a single gene. One example is cystic fibrosis that affects around one person in every 10,000. Huntingdon's disease, sickle cell anaemia and Tay-Sachs disease are others. These are your parents' fault in the sense that it is usually their genes that have passed the disease onto their offspring, although it's a bit tough to blame them.

The genetic mutations are clearly understood in an increasing number of these diseases. These include cystic fibrosis, sickle cell disease and haemophilia. The examples are given to illustrate the progress that has been made in characterising some of these inherited diseases.

Cystic fibrosis – Autosomal recessive inheritance. Caused by mutation in the gene cystic fibrosis transmembrane conductance regulator (CFTR) on chromosome 7. The CFTR gene is some 250,000 base pairs long and creates a protein that consists of 1,480 amino acids. The commonest cause of cystic fibrosis is the absence of just three DNA nucleotides on the gene leading to a deletion of phenylalanine at position 508 on the protein.

Sickle cell disease – Autosomal recessive inheritance. Caused by a mutation in the gene ß globin that results in glutamine being substituted with valine on position 6 of the ß-haemoglobin chain. The defective gene is on chromosome 11.

Haemophilia A (there is also haemophilia B) – X linked recessive inheritance. This is caused by several gene abnormalities, including inversion of intron 22 in the factor VIII gene at position 28 on the long arm of the X chromosome. An inversion occurs when a chromosome is reversed end to end. An intron is part of a gene that does not directly code for a protein.

Down's syndrome – the most common cause is an extra chromosome, number 21, so there are three instead of the normal two chromosomes. This is caused by an error in cell division and results in learning and developmental disabilities. In most cases, it is not inherited and is found in all races and all countries and usually occurs because of a chance happening at the time of conception

Not all neonatal problems are caused by faulty genes. If a mother is infected with Zika virus it can affect foetal development so that the child is born with an underdeveloped head and brain. Other infections such as rubella or toxoplasmosis can also cause brain damage. Factors that may cause genetic defects or affect neonatal development include lack of folic acid (associated with spina bifida), drugs such as thalidomide, smoking, alcohol and drug abuse and malnutrition. In addition, problems during delivery may be fatal or result in a seriously damaged baby. These include the umbilical cord being obstructed, massive maternal blood loss or obstructed labour. Cerebral palsy is a common outcome in these cases.

Of all the genetically determined diseases the ones that most directly affect the ageing process are called progeria syndromes. In these, ageing is accelerated with cells becoming old much earlier than usual and individuals showing the signs of ageing in their teens or early twenties before dying. They are a group of rare genetic disorders causing those who have inherited one of them to appear to age rapidly and prematurely.

Many different manifestations of these syndromes have been identified but they can be broadly divided into two categories based upon the underlying mechanism. Some are caused by abnormalities in the envelope surrounding the nucleus (e.g. Hutchinson-Gilford Progeria [HGP]). Many other progeroid syndromes are caused by defects in DNA repair. Mutations in proteins that repair breaks during DNA replication are known to cause Werner, Bloom and Rothmund-Thomson syndromes. Defects in other DNA repair proteins are the cause of other syndromes including Cockayne and Seckel syndromes.

HGP is the commonest and most widely studied of these syndromes. It is caused by a point mutation in the LMNA gene. This gene provides instructions for making proteins called lamins that are essential for supporting the membrane surrounding the cell's nucleus. They may also have a role in organising the genome. Instead of lamin A being produced, a truncated version called progerin is the result. It appears that a simple replacement of cytosine (C) by thymine (T) at nucleotide 1824 of the LMNA gene causes this catastrophic outcome.

People with progeria tend to have small, fragile bodies. Their skin is thin and wrinkled and they get diseases, such as atherosclerosis, osteoporosis and kidney failure, usually associated with old age. Although their bodies show outward signs of ageing their minds are usually unaffected. Because carriers rarely live long enough to reproduce it probably arises from a spontaneous mutation.

Cells from patients with progeroid syndromes have been grown and developed into pluripotent stem cells. This has provided a model for examining the molecular mechanisms of ageing. Mice have also been engineered to have features of the progeroid syndromes providing another way of studying this problem.

The defining features of progeroid syndrome are as follows:
Increased DNA damage and defective DNA repair
Telomere dysfunction
Epigenetic changes and aberrant gene expression
Defects in the nuclear envelope
Defects in cell growth and division
Cellular senescence – loss of capacity of the cell to grow and replicate
Metabolic defects including insulin resistance and decline in growth hormone values
Chronic inflammation
Stem cell exhaustion with a fall in the capacity to regenerate tissues

Although progeria looks like ageing in many respects, it is not ageing. But understanding these syndromes provides valuable insights into the mechanisms that might underlie normal ageing. The same molecular pathway responsible for Hutchinson-Gilford Progeria is active in healthy cells, and older people acquire cellular defects similar to those seen in progeria. These include changes in histone patterns and increased DNA damage. Lamin A metabolism is seen as a potentially important target for anti-ageing therapy. It is also to the testament to the complexity and effectiveness of the properly functioning human machine, and how little it takes to degrade and destroy its working. For those who have been unfortunate to have inherited progeria there is also the hope that interventions will be found that will ameliorate or even cure the condition.

A single gene found at the same place on the chromosome can commonly have several alternative forms or alleles. Several individual alleles have been shown to be associated with longevity in humans. For example, there is a specific gene linked to smoking fewer cigarettes.

It is also strongly associated with longevity, perhaps because of the effect on smoking behaviour. Longevity is also associated with alleles that are relatively protective for several cardiovascular diseases. Some of the risks associated with longevity genes are manageable with lifestyle and dietary changes. These include lipid levels, high blood pressure and obesity. Together these results suggest that ageing is multifactorial, but that interventions to prevent or manage chronic diseases may be important measures when seeking to extend the duration of life.

There has been speculation that some of the responsible genes may be suitable for drug development. Information on these is being added to and made freely available. As of 2013 there were 246 published studies looking at 751 different genes and 1987 variants. There are thought to be 300 to 750 genes, or perhaps more, that affect human longevity. With ageing there is a change in gene expression. One study identified 33 genes that under-expressed and 21 that over-expressed in brains of the elderly. Other work has revealed 56 consistently over-expressed and 17 under-expressed genes in the old. Genes thought to be relevant include those involved in the insulin signalling pathway, lipid metabolism, DNA repair and RNA regulation. By comparison it has been reported that 697 variants may influence human height. The genetic contribution to longevity is complex and any genetic modification is likely to be difficult and fraught with possible unexpected ramifications.

Genetic damage accumulates throughout life. External physical, chemical and biological threats challenge DNA integrity. Over time the incidence of DNA replication errors becomes more likely, and oxygen free radicals cause damage. Errors arise including mutations in single nucleotides, translocation of genes, chromosomal gains and loses, telomere shortening and gene disruption caused by viruses. DNA repair mechanisms become less robust with the net effect that old age is associated with increased DNA damage.

Human mitochondrial DNA has 16,569 base pairs containing 37 genes coding for 24 different RNAs and 13 polypeptides. Over time, point mutations arise as well as rearrangement of larger segments of the genome. These abnormalities accumulate faster in mitochondrial DNA than in the main cellular DNA in the nucleus. The impact of these changes is hard to quantify and may or may not be relevant. Mitochondria are key elements in the proper functioning and health of cells. As with many age-associated changes it is, as yet, unclear whether the observed changes contribute to the ageing process or are simply a reflection of the stresses that occur over time.

Within the DNA in older people there is something called methylation drift. As the name implies this is not directional and is quite variable and patchy. This argues against it being a totally programmed change in gene expression or an age-related change in writers or editors of the methylation code. There is greater drift in rapidly dividing tissues suggesting that there may be inaccuracies in the copying of DNA methylation patterns. Chronic inflammation, such as experienced with some bowel diseases or seen in smokers, also appears to modify methylation. And there is even speculation that dietary influences, including high levels of folic acid or vitamin B12, may have an effect. The pattern is confusing in older cells with some methylation sites showing increased activity and others showing less than expected. Centenarians have fewer age-related methylation changes and may be able to pass on this characteristic to their children. Rhesus monkeys and mice exposed to 30 per cent to 40 per cent calorie restriction had less age-related methylation drift compared with controls with free access to food. The authors suggest that provides further evidence of the importance of epigenetic drift as a determinant of lifespan in mammals.

Queen bees may live for years but workers with identical genes have a life span measured in a few weeks. They look different and have different roles. The answer also lies in the degree of gene expression and is driven by nutritional differences while in the larval stage. Cancer cells have a high degree of aberrant DNA methylation as well as other epigenetic alterations. Epigenetic changes are thought to be important reasons for why cancer occurs and for this reason are important contributors to mortality.

It has been found that methylation often occurs in places where a cytosine precedes a guanine (CpG) in the DNA. One study found that methylation at CpG sites has a correlation of 96 per cent with tissue age. Not only could it identify age but also it revealed that the rate of ageing changed over time. In the early years the rate was high before decreasing to a slower linear pattern once adulthood was reached. In some of the CpG sites there was decreased methylation and in others increased methylation. This finding has been replicated in several other studies showing that the association of methylation-based biomarkers did not differ significantly across subgroups of race, ethnicity, gender, body mass index (BMI), physical activity, smoking history or major chronic disease.

Individuals whose epigenetic age is greater than expected for their chronological age are at an increased risk of dying. On the other hand the children of supercentenarians (in this study those who lived to between 105 and 109 years) have a lower epigenetic age than matched controls. If this can be confirmed it suggests that epigenetic age captures some aspect of biological age and with it susceptibility to disease and risk of dying. It is an enormous step from this finding to suggest that intervening to alter the pattern of methylation will affect chronic diseases and longevity. It is, however, a worthwhile avenue to explore. Some other measures of ageing are related to the number of times a cell replicates. The epigenetic technique is interesting as it tracks age in cells such as neurons that do not proliferate, as well as in so-called immortal cancer cells.

Already several uses are being made of this discovery. Criminal investigators are looking at whether biological residues left at a crime scene may help pin down the age of the perpetrator. It may be relevant where there are discrepancies between a person's epigenetic and chronological age.

In other words if your tissues are ageing faster than the clock says you should, is this a marker of disease or other medical risk?

So far at least one study has shown that an increase in epigenetic age is associated with an increased risk of dying. Distortions in epigenetic age also seem to be relevant for some cancers, in the morbidly obese or those with HIV infection.

Perhaps the more interesting question, at least with respect to ageing, is the underlying biological process. Is epigenetic ageing involved in causing the ageing process or is just of reflection of the passage of time? Methylation can, in theory, be altered. If it is somehow programmed then it might be amenable to being halted or even reversed with a measurable impact on health and longevity. On the other hand it may simply be a marker of the passage of time, a cellular record of the accumulation of epigenetic errors in the way that lines and furrows on a face bear testament to the years that have been lived.

With the present stage of knowledge most of the studies are fishing expeditions. They look at associations between genes and outcomes such as longevity, cancer and age-related diseases. The mechanisms are complex and poorly understood and influenced by many factors. To try and make sense out of this will be a mammoth task. From all the potential interventions, finding those that are truly beneficial in humans is likely to be extraordinarily difficult. There may be key pathways that can be modified to give a consistent therapeutic effect. But somehow I doubt it. If this case any progress is likely to be slow and incremental.

13 Ageing: Other organisms

"For the fate of the sons of men and the fate of beasts is the same. As one dies so dies the other; indeed, they all have the same breath and there is no advantage for man over beast, for all is vanity. All go to the same place. All came from the dust and all return to the dust."
New American Standard Bible, 1971, Ecclesiastes 3:19 and 3:20

Ageing is common to almost all creatures, but not all. *Turritopsis nutricula* is a small jellyfish that may be immortal. From the mature stage it can revert to the polyp phase thus rebooted to start a life cycle all over again. *Hydra vulgaris* fresh-water polyps do not show evidence of ageing. Senescence is not seen and stem cells continually self-renew. Hydra therefore escape ageing through a process of constant regeneration. Some human cells are immortal. The HeLa cervical cancer cells are one example. Some animals hardly age at all. Closely related animals become old at very different ages. Others, such as the salmon, age precisely in time with a biological clock. A mouse will be old at two years of age, a dog at 12, a cat at 16 and an elephant at 40. In many ways, it is extraordinary that there is so much difference between the lifespans of different mammals, while all the animals in the same species are fated to similar outcomes.

There is clearly a genetic element associated with ageing. Some creatures are programmed to die after a single reproductive episode. This is called **semelparity**, colloquially referred to as "big bang" reproduction. This strategy means that all available resources can be devoted to maximising reproduction in a trade off with further years of life. Examples of semelparous creatures include the octopus, Pacific salmon, many insects and annual plants.

Iteroparity is the ability of an organism to be able to reproduce multiple times over its lifetime.

Mutations in genes affecting endocrine signalling, stress response, metabolism and telomere length have all been shown to dramatically increase the lifespans of some organisms. The insulin/IGF-1 pathway influences life span in worms, flies and mammals. *C elegans* roundworms with mutations affecting this pathway live twice as long. Study of this pathway has conclusively shown that the ageing process, at least in the roundworm, is subject to regulatory control. Manipulations of environment and genes have now managed to increase the worm lifespan tenfold. Mutations in this pathway have also been shown to increase lifespan in mice. In the fruit fly (*Drosophila melanogaster*) the gene CHICO increases lifespan by nearly 50 per cent. Impaired IGF-1 receptor activity has even been linked to centenarianism in Ashkenazi Jews, and other variants appear to confer exceptional longevity in several human groups.

Many of these long-lived mutants are resistant to age-related illnesses including cancer and heart disease once again raising the possibility that a range of diseases of old age may be treated by focusing on the ageing process.

In most mammals and songbirds function deteriorates markedly with age. In contrast long-lived seabirds seem not to have much in the way of physiological deterioration or loss of reproductive ability, until close to the end of their lives. Maybe in order to survive these birds must be able to fly to feed, migrate and evade predators throughout their lives. Although birds have a high rate of metabolism, on the whole they live longer than mammals of comparable body size.

The record lifespan in years for some animals is reported to be:

House mouse	4
Norway rat	7
Dog	29
Cat	38
Polar bear	42
Horse	62
Asian elephant	86

Macaws can live up to 100 years in captivity, Koi carp for up to 200 years, tortoises 190 years and whales more than 200 years. Greenland sharks are the longest-lived vertebrates with a lifespan of at least 272 years and possibly up to 400 years.

Longevity

What lessons can be gleaned from other living organisms? AnAge, the animal ageing and longevity database, provides interesting information about creatures with extreme longevity. In particular, they feature eight species that show few signs of ageing. These are:

Species	Recorded longevity (years)
Rougheye rockfish	205
Olm (proteus or cave salamander)*	102
Painted turtle	61
Blanding's turtle	77
Eastern box turtle	138
Red sea urchin	200
Ocean quahog clam	507
Great Basin bristlecone pine	4,713 or older

***The Olm is a blind amphibian that lives its whole life in darkness**

"Clamgate" was the banner headline in the press after the death of Ming, the quahog clam. This story made news round the world in 2013. The Welsh scientists involved were accused of being clam murderers for killing an ancient clam. In fact it was dredged up from the Icelandic shelf and then frozen. It was only later by counting the number of bands in its shell that age could be estimated.

The naked mole rat (NMR) is an extremely long-lived mammal that has been extensively studied. Although the size of a mouse its maximum lifespan is more than 30 years. When examined they show few signs of getting old, they maintain high fecundity until they die and have an extremely low risk of developing cancer. They live in large colonies with a single breeding queen who suppresses the sexual maturity of her potential rivals. They live in complete darkness in an environment with low oxygen (about eight per cent) and high carbon dioxide (>20 per cent) concentrations. They can even survive 18 minutes of total oxygen deprivation by switching to anaerobic metabolism fuelled by fructose. At a cellular level there is little sign of stress-induced damage.

The genome of a NMR has been analysed to look for insights into its remarkable physiology. It shows that NMRs split from their cousins, rats and mice approximately, 73 million years ago. A small number of genes (some 45) have been positively selected over time. Among them are genes affecting telomere function showing greater stability during ageing. They have unusual ribosomes, the machinery that converts genetic instructions into proteins. This may be why they show more accurate protein translation compared with mice. There is little change in gene expression between young and old adults, something that is quite different from that seen in humans and other mammals. Compared with a mouse the gene mutation rate in a NMR is much lower and there is a higher expression of DNA repair genes. The relative stability in the expression of proteins could be one reason why they live so long. As NMRs live in complete darkness they do not synthesise melatonin, the hormone that normally modulates sleep and circadian rhythms in response to sunlight and darkness. However, the genes involved in melatonin synthesis remain intact although gene expression is very low. Many of the genes associated with vision are missing. Adaptations also exist to cope with the low oxygen environment. The NMR is relatively insensitive to pain lacking the mediator Substance P in its skin and associated pain fibres. This may be an important factor related to longevity as mice without pain receptors live longer than controls.

In summary, despite its unusual longevity and lack of senescence, there is nothing extraordinarily different about its genome compared with other mammals. The differences that do exist suggest that relatively minor genomic changes might substantially alter outcome.

Tardigrades, or water bears, space bears or moss piglets, may be the toughest animals on the planet.

They are eight-legged, water-dwelling micro-animals, about 0.5mm long. They are also almost indestructible surviving desiccation, huge extremes of temperature (from minus 273⁰C to 100⁰C), high doses of radiation and even the vacuum of space. Their genome has been sequenced and contains a few surprises. It has 16 copies of antioxidant enzymes while most animals have fewer than 10. They also have four copies of DNA repair genes compared to the single gene most animal cells express. Several oxidative enzymes usually found in other animals were missing. The water bears produce a protein called Dsup that suppresses radiation damage. When incorporated into human cells in culture it reduces radiation damage to DNA by 40 per cent. In common with many of these studies it raises more questions than it answers while demonstrating that the same biological mechanisms can confer a wide range of properties in living creatures.

Humans and other mammals have a considerable capacity to regenerate tissues. A foetus in the womb can regrow a complete fingertip. After birth a child's wound will heal rapidly without scarring, and broken bones will remodel. With increasing age human regenerative capacity falls. For example, in mammals hair cells in the inner ear are essential for hearing. They can only be regenerated during early development. In salamanders, birds and fishes hair cells can regrow throughout adulthood.

Across the animal kingdom there is a huge range in the ability to regenerate tissues and organs usually through the mobilisation and activation of stem cells. Some vertebrates such as zebrafish and salamanders can regrow complex structures including limbs, heart, brain and spinal cord. Mammals rely primarily on tissue-specific stem cells for their regenerative capacity. As we get older these cells lose much of their capacity to proliferate and renew. This has been shown in muscle, blood components, bone, brain and other tissues.

So why can some organisms have extensive regenerative abilities, and why do they not lose this ability with age?

The most extreme examples usually quoted are the *hydra* and planarian flatworm, both of which can regenerate whole bodies from a single fragment. Even zebrafish can continue to regrow their caudal fin, barbels and heart into old age. Experiments show that new heart muscle is, in fact, not based on stem cells but rather from activation and proliferation of spared cardiac muscle cells. The hormone leptin b is induced in regenerating hearts and fins of zebrafish and may trigger gene expression in injury sites leading to regeneration. A chemical soup of mediators has been identified which control regenerative growth.

Making sense of it all is a different problem.

Salamanders, such as newts and axolotls, continue to regenerate whole limbs as an adult. Several cellular and molecular mechanisms are associated with exceptional regenerative capacity. Failure of these mechanisms is often seen with ageing. A sophisticated DNA response is required to detect and repair damage and maintain genomic stability. If the damage is too great cells are removed, commonly through apoptosis. In particular a protein called p53, sometimes referred to as 'the guardian of the genome', may play a central role in protecting DNA by promoting cell cycle arrest, apoptosis and senescence. p53 works by binding to DNA and activating the expression of several other genes. It has an important role in preventing cancer and as many as 50 per cent of all human cancers contain mutant p53. Boosting normal p53 concentrations is being looked at as a cancer treatment. In mice, high levels of p53 lead to ageing and earlier death. A decrease in p53 activity is necessary if salamanders are to regenerate limbs. These findings suggest that increases in p53 prevent regeneration.

Telomere length also has something to do with regeneration. High telomerase activity is found in animals with good regenerative capacity whereas shortened telomeres are seen in those with poor ability. As with ageing, mitochondrial preservation and epigenetic stability have also been implicated.

More than anything the study of other organisms provides hope that ageing may be treated and regeneration could be possible. If other animals can repair themselves and live forever then why can't we? There is an awfully long way to go before we fully understand how to prevent ageing and to promote regeneration, and it is a big step from there to making it work in humans. It might be easy to start with something simple like scalp hair growth. Genes that may cause baldness have been identified and an increase in prostaglandin D2 may be the mechanism for lack of hair growth.

Scientists have already cloned a new animal from a single cell. It won't help me much if I'm regrown as a copy. It would be nice though to bud off a replacement arm or leg if something happens to the ones I've got.

The huge natural age range of similar creatures and the potential for some animals to live extraordinary long lives suggests that there are factors that can be identified and modified to stimulate regeneration and to extend human ageing.

14 Lifespan: Effects of lifestyle

"When you go to war as a boy, you have a great illusion of immortality. Other people get killed, not you... Then, when you are badly wounded the first time, you lose that illusion and you know it can happen to you."
Ernest Hemingway (*Men At War,* 1942)

Assuming you've inherited good genes and have been born undamaged, what else can affect your chances of a long and healthy life? The body and mind are an organic machine and steps must be taken not to damage it. In England and Wales official statistics say that those who die early from avoidable causes lose an average of 23 potential years of life. For children and young adults accidents and the complications of childbirth are the commonest reasons. If you want to take part in base-jumping, heli-skiing, cave diving or eventing, you will increase your chances of having an interesting, but possibly shortened, life. For the older adults cancer and heart disease mainly cause early avoidable deaths.

There are ways of abusing your body that will take their toll over many years. It is obvious that if you smoke tobacco the chances of achieving a long life are considerably lessened. It causes more preventable deaths than anything else. Smoking will, on average, knock about 10 years off life expectancy. It is the primary cause of 30 per cent of all cancer deaths and 80 per cent of deaths from chronic obstructive pulmonary disease (COPD). The good (ish) news is that if you give up by your mid-30s then you can over time more or less catch up with those who never smoked.

Habitual drug and alcohol abuse will also eventually take their toll, although the number of old rockers still going strong attests to the recuperative power of the human body. In the United States, alcohol abuse accounts for between four per cent and 10 per cent of premature deaths. The amount of alcohol consumed and the pattern of consumption are associated with different health outcomes. For example, heavy episodic drinking is associated with sudden cardiac death and injuries, whereas the risk of cancer is related to the total amount consumed over time. Some reports suggest that drinking a moderate to modest amount of alcohol may protect against cardiac disease, although other interpretations of the evidence would dispute this. Wine drinking appears to be associated with lower risks compared with drinking beer or spirits. However, this may simply be a reflection of the characteristics, lifestyle and diet of the drinkers.

It has been estimated that some seven per cent of the population of the United States – about 23 million adults – are alcoholics. Alcoholics Anonymous does not specify a number of units of alcohol to define this. Rather it can be described as a physical compulsion coupled to a mental obsession. Alcoholics not only have shortened lifespan but heavy drinking often disrupts their work and social life. A unit of alcohol is defined as 10ml of pure alcohol. This equates to a single measure of whisky, one-third of a pint of beer or half of a standard (175ml) glass of wine.

The 2016 guidelines from Prof. Dame Sally Davies, the UK Chief Medical Officer, state that the risk of developing a range of illnesses, including cancers of mouth, throat and breast, increases with any amount drunk on a regular basis. She recommends not regularly drinking more than 14 units per week, and to spread drinking out over several days rather than binge.

As fewer people smoke tobacco, obesity is becoming the greatest behavioural health risk. It is an over-simplification to say that people get fat because they eat too much and exercise too little. That's the sort of phrase that is likely to stir up a storm on the Twittersphere.

In mid-Victorian England an adult would consume 3,500 to 4,000 calories a day in contrast to the 2,000 to 2,500 recommended today. This was justified by the higher rates of activity necessary to live in those times. Over my lifetime obesity has, inevitably, become a big problem. In hospitals we don't use the word fat, or even obese. Bariatric is the politically correct word to use. Bariatric surgery is increasingly common. Surgical techniques are used to restrict stomach size so patients can't eat as much. I'm not claiming it's easy to cut down on food and considerable responsibility falls on the food industry producing processed food containing unwanted salt, sugar and fats. The metabolism of individuals does vary. However, with a few exceptions for patients with rare endocrine disorders, if less food is consumed and more exercise taken, weight will be lost. Going without food every now and again, in other words periodic fasting, may have additional benefits. Not only will it reduce body weight and fat, but it has been shown to lower blood pressure, and decrease circulating values of IGF-1, a protein that has been implicated in ageing.

Most studies examining the impact of food on weight rely on patients recording their dietary intake in diaries. Some studies suggest that calorie intake has fallen compared with values seen in the 1970s, despite rising obesity levels. In a national survey many people even claimed to be eating less than was necessary to stay alive. These claims needed to be questioned. A study from the UK Office of National Statistics (ONS) has looked at the accuracy of this self-reported data. Doubly-labelled water measures energy expenditure and provides an objective measure of calorie intake. This technique was used to examine the differences between what participants said they ate and their actual energy intake. Unsurprisingly, perhaps, participants under-reported their calorie intake by some 20 per cent to 35 per cent. It may be wise to bear this in mind when thinking about any study that looks at the association between nutrition and outcomes.

In a book on longevity, obesity is easy to deal with. Obese people will, usually, die younger. BMI (body mass index) is often used to quantify obesity, although waist circumference or waist hip ratio are also relevant and may be better. Worldwide the incidence of obesity has more than doubled since 1980 with 13 per cent of all adults estimated to be obese, defined as a BMI greater than or equal to 30 (BMI is calculated as body weight in kilograms divided by the square of height in metres. Thus the BMI of someone who weighs 80kg and is 1.8m tall is 24.7).

In the United States about a third of the population are obese. The United Kingdom is not far behind with a quarter of adults being obese, four times as many as 30 years ago.

The range of normal BMI is between 18.5 and 24.9. Because obesity has become so common it needs to be further subdivided with those with a BMI of less than 35 being Class I obese, those between 35 and 39.9 Class II, and those with BMIs of 40 and greater Class III or extremely obese. The term morbidly obese refers to all those with a BMI of 40 or more as well as those greater than 35 who have obesity-associated health problems such as high blood pressure or diabetes. Class III obesity affects some six per cent of adults in the United States so it is in both senses not a small problem. Unsurprisingly, this degree of obesity causes more deaths, primarily from heart disease and cancer, and the higher the BMI the greater the incidence of early mortality. Compared with those with normal BMIs, subjects with a BMI between 40 and 45 died on average six to seven years sooner rising to some 14 years for those weighing in with BMIs in the 55 to 60 range.

Eleven cancers associated with obesity
Oesophageal adenocarcinoma
Multiple myeloma
Lower part of stomach (cardia)
Colon
Rectum
Biliary tract (liver, gall bladder, bile duct)
Pancreas
Breast
Endometrium
Ovary
Kidney

Interestingly, evidence suggests that being slightly overweight (BMI of between 25 to 30) may be a good idea. There does not appear to be any strong evidence of harm and there may even be benefits in terms of recovery from critical illness, after major surgery or following heart attacks. This is sometimes referred to as the obesity paradox. Fat may have a protective effect and the energy it stores may help overcome trauma and disease. Unlike those who weigh less, obese people also have ways to change their destiny. They can certainly modify their lifestyle, weight and eating habits. There is also evidence that underweight patients have worse outcomes in a variety of medical conditions that cannot be explained by the effect of the disease causing weight loss.

If a thin person has a heart attack it's likely to be from bad genes and there's not much that can be done to decrease the risk. Perhaps thin people are more susceptible and have fewer reserves.

Drinks containing artificially sweetener or sugar are associated with a higher risk of neurological problems including stroke and dementia. The Coca-Cola Company sells more than 1.9 billion drinks worldwide each day and on the basis of current evidence I'll continue to have the occasional Diet Coke. The evidence is based on relatively small numbers and an association (rather than causation) provides feeble justification to support a major lifestyle change. In the meantime be reassured that it is the sweetener not the bubbles that is being called into question, and you can continue to drink sparkling water and champagne (in moderation, of course).

A sedentary lifestyle is also something to avoid. Not only does physical activity contribute to a longer life but will also delay or prevent the onset of several conditions including diabetes, hypertension, colon cancer and depression. Amazingly it has been reported that the average American walks less than 3,000 steps per day, equivalent to walking only about a mile and a half. The relative risk of death is some 25 per cent lower in active and fit people compared with the rest of the human race.

Although the benefits of exercise are uncontroversial not that much is known about how it works. Various compounds are somewhat lower in those people who exercise. These compounds include triglycerides, high-density lipoproteins and tissue plasminogen activators. It is also known that the cells of those who regularly exercise are more resistant to oxidative stress. Exercise can alter DNA methylation patterns. This has been shown to lower the expression of genes that can cause cancer and to increase the expression of genes that may suppress it.

A prospective study of more than 2,000 healthy elderly men found that regular exercise was associated with a 30 per cent lower risk of dying before 90 as well as significantly better physical function. If at 70 a man didn't smoke, was not obese, took exercise and did not have hypertension or diabetes they had a 54 per cent chance of getting to 90. If they had two of these adverse factors the chances were between

22 per cent and 36 per cent. However, if they had all five they only had a four per cent chance of surviving another 20 years. Another study followed up runners older than 50. After 20 years the runners had far less disability than non-running controls, and half the risk of dying. Regular short bursts of tough aerobic exercise (high-intensity interval training [HIT]) may be the best way to exercise, particularly in the elderly. It has been shown to reverse many age-related effects on protein expression and improves muscle mitochondrial activity.

When it comes to failing brain function a large Finnish study showed that a structured two-year intervention of diet, exercise and cognitive training improved or maintained cognition or perception.

Sleep is also important to well-being. Several studies have looked at the association of the amount of sleep and mortality. Short sleep duration is typically defined as less than six or seven hours per night. Long duration is usually more than eight or nine hours a night. About half of all people report getting a bad night's sleep so this is an important question. Analysis of the data leads to the conclusion that too little and too much sleep are both associated with increased mortality. This is, of course, an association not a causal relationship, and is not evidence that the duration of sleep actually causes the problem. Some studies have shown that the amount of sleep we need is genetically influenced. If so, whether you don't need much sleep, or require a siesta every afternoon, may not be a matter of personal choice. A particular gene has been found in a family of natural short sleepers. When mice were engineered to express this gene they seemed to need less sleep. Wouldn't it be wonderful to sleep less without feeling tired or having serious side effects? Perhaps it will be possible in the future.

You will probably not be surprised to learn that the better educated and wealthy also live longer. This may have something to do with better diet, better housing and better understanding of what living a healthy life entails. It is probably the differences in lifestyle that explain why the premature death rate in Scotland is consistently higher than that in England and Wales.

Happy people live longer and there is reason to believe that this is a genuine effect. Of course, people who are unwell may be unhappy and it is difficult to be sure of the effect of mood on well-being. It also helps to be married, to be sociable and to keep mentally and physically occupied. The evidence is strong and consistent. If you smoke, are obese, drink to excess, and don't take exercise then you should not be surprised if you become unwell, suffer disability and die earlier than those who look after themselves.

Having got the basics out of the way is there anything that can reasonably be done by way of lifestyle to maximise lifespan? Given the widespread potential of many people to get to old age I'm naturally a bit suspicious of anyone who claims to have found the secret among a group of wizened old people living in a hidden valley. That, more or less, is the hypothesis underlying the so-called Blue Zones. Researchers identified a few areas in the world where clusters of extremely old people live.

They are:
Barbagia region of Sardinia, Italy
Acciaroli, Italy
Okinawa, Japan
Seventh Day Adventists in Loma Linda, California
Nicoya Peninsula, Costa Rica
Icaria, Greece
Öland, Sweden

The next step was to look at what sort of lifestyles these old people have to see if there is anything that may be responsible for their longevity. Unsurprisingly, in these relatively poor, remote and often hilly places, people walk a lot, eat simply and make their own entertainment within the community. They have little opportunity to get fat or laze around.

A recipe for long life can thus be formulated, and indeed turned into an industry catering for those who need a guru to tell them how to live. I'm not knocking it. As a lifestyle there is a lot in it to commend, and anyway it more or less agrees with the message of the wider epidemiological studies.

> **Nine things common to all these areas**
> Activity as a natural part of daily life
> Having a sense of purpose
> Taking some time to reflect and de-stress
> Not over-eating
> A diet without much meat in it
> Moderate alcohol consumption
> Belonging to a community
> Being part of a family
> Having a sociable life

However, it is important to remember two things. Firstly an association between lifestyle and outcome does not imply causality. Just because these people live a long time and have a certain way of living it does not mean that one is the cause of the other. No one would suggest the rise in stockmarket indices or car ownership was a cause of increased longevity in the general population, but they have both increased over time. The second thing is that these people may have different genes. A small isolated community is an ideal setting for a little bit of inbreeding, and lusty older men taking young brides may just provide a selective pressure towards longer life.

One study comparing really long-lived people with their shorter lived contemporaries found that centenarians were as likely to be have been overweight or obese, to have smoked, drank alcohol, not exercised or not followed a low calorie diet. This strongly suggests that survival to be 100 or more has much more to do with genes than lifestyle. The answer should become clearer over time. If those in the Blue Zones become wealthier and less isolated their way of life will change. They may take less exercise, change their eating habits and engage less frequently in social and community activities. On the other hand, advocates of this way of life may demonstrate the benefits. Let's wait and see.

I love a rogue, particularly an old rogue. Past a certain age inhibitions can diminish and some old people say the most outrageous things and get away with it. What I particularly love is when a centenarian confounds expectations. Advice may be to live sensibly, eat a Mediterranean diet, take exercise and do everything in moderation. But there are exceptions; hard drinking, womanising, oldies who have disobeyed the rules and, despite all that, continue to live a long and healthy life. A review of the lifestyle of Polish centenarians found that four out of 336 still smoked and about 10 per cent had actively smoked in their lifetimes with an average of 28 pack years (or smoking 20 cigarettes a day for 28 years). It is possible to grow old disgracefully while disobeying the advice, it's just a lot less likely.

It is said that life isn't fair. What consolation is there for those who do all the right things and succumb from cancer or heart problems far too early? What merit is there in carrying on into a dotage in a small, boring, careful life when out there are geriatrics dancing on tables and being the life and soul of the party? How annoying that must be to those who have striven to do the right thing all their lives.

Life is to be enjoyed not just prolonged.

In the more racy publications there are many case histories of individuals who have beaten the odds. They are the exceptions that prove the rule. They are not actually beating the odds, they are the odds. The chances of winning the UK lottery jackpot are 1 in 45 million. Buying a ticket only gives slightly better odds than not having one at all. However almost every week someone wins. The odds only apply to an individual ticket. If 45 million tickets are bought then the odds of one of those tickets winning become quite good. When it comes to predictions of the effect of lifestyle choices on outcome it is a similar statistical phenomenon. An association is noted between early death and smoking, lack of exercise or diet. All these things increase your chances of dying early but do not make it inevitable. Out of several million people there will be a few who are the ones who survive against the odds. However, if you stack the odds too much they cannot be beaten.

I wouldn't put much money on a morbidly obese, hypertensive, smoking, alcoholic, diabetic living long enough to draw their state pension.

I really like these examples as they appeal to my iconoclastic nature. However, a word of caution, these are newspaper reports and do not necessarily have to be believed.

Dorothy Howe on her 100th birthday attributed longevity to Bells Whisky and Superking Black cigarettes.

Jeanne Calment who lived to 122. She loved chocolate and port wine and only gave up smoking in the last 5 years of life.

Paulina Spagnola is reputed to have given her secret for longevity as lots of booze

104-year-old **Elizabeth Sullivan** put it down to drinking three sugar-loaded Dr Peppers a day.

Pearl Contrell, aged 105, ate bacon every day.

Tom Spittle, 100 in June 2015, had an English fry-up breakfast and a pint or two in the evening. He enjoyed gambling and smoking.

Lizzie Seymour, died aged 117 years on April 15, 2017. She was the last living person known to have been born in the 19th century. She attributed her long life to eating three eggs a day, two of them raw, and to kicking out her husband in 1938. Reports said that she rarely ate vegetables or fruit.

15 Lifespan: Maintenance and repair

"Since we have had a history, men have pursued an ideal of immortality
George Wald (American scientist, 1906-1997)

Let's imagine your body is a motor car. If you want the car to last a long time there are a few obvious things that would help. First of all buy one with a reputation for excellent design and build quality. Today most cars are reliable, don't rust and rarely break down. However, few people would expect their REVA G-Wiz (worst car ever according to the magazine *Auto Express*) to last 10 years whereas a 2018 model Rolls Royce Ghost will probably still be on the road well beyond 2050. It clearly helps where the car is kept. An air-conditioned, temperature-controlled, secure underground garage is going to preserve a vehicle longer than parking it outside on a windswept seafront road in a cold and damp climate. The car should be cherished, driven sensibly and the mileage kept low. Crashes must be avoided although sometimes they happen no matter what steps you take. Obviously regular preventative maintenance is important. Servicing must be carried out according to the manufacturers schedule with recommended consumables and replacement parts. If repairs are needed they need to be done by reputable mechanics in a licensed centre using genuine spares.

All of us have the body that fate dealt out to us and some inhabit G-Wizes and others Rolls Royces. Clean water, adequate nutrition, effective sanitation, good quality housing, clothing and heating enable almost everyone to have a chance at a long life. Medical progress is the icing on the cake once all the basic needs have been satisfied. Whatever body we are born with we make choices about what to do with it. In the last chapter I wrote about the harm that can be done by abusing your body. I'm going to now consider the ways in which conventional medical science helps us live healthy and long lives.

Looking back makes me realise how recent are the developments in medicine that I took to be routine during my working lifetime.

The story of anaesthesia is fascinating. Ether was known about from the 13th century and its hypnotic effects noted in 1540. Known to have mind-altering properties, it was used as a recreational drug in the 1800s. Anaesthesia could have been given for three centuries or more before its successful demonstration on October 16, 1846 in the Massachusetts General Hospital, Boston, Massachusetts, USA.

Anaesthesia was one of those unknown unknowns that came along and changed the world. Conquering pain removed one of mankind's greatest fears, made possible modern surgery and may have changed attitudes to everything from child labour to the rights of the aristocracy.

For if God-given pain could be relieved it was possible to question the place of everything in the world. Word of its discovery swept throughout the world in a matter of weeks, as fast as steamships could convey the details. Almost everywhere ether was being administered within days of arrival of the news.

Throughout my career I never lost my wonder and awe at what anaesthesia can do. Routinely I gave chemicals that sent someone to sleep. I watched over them while a surgeon rummaged inside their abdomen, or exposed their brain, or replumbed the blood vessels of the heart. Then I would wake my patient up and they would have no recall of what had happened and no long-term effects (and very few short term effects) from the anaesthetic. In an extraordinary leap of faith I assumed the patient would wake up, and they always did. In the intensive care unit we could sedate people for days or weeks, taking away pain and anxiety, while our efforts treated the patient and allowed time for their bodies to heal. During the more than 30 years of my career in clinical medicine everything changed. None of the monitors and none of the drugs that I used at the end of my career existed when I gave my first anaesthetic.

Although anaesthesia was essential for safe surgery it also needed asepsis, pioneered by Joseph Lister in 1865 with the use of carbolic acid, to reduce surgical infections. The first gall bladder operation was in 1867 and abdominal surgery only became routine, although still hazardous, at the start of the 20th century. The first successful lung operation was performed in 1933 and open-heart surgery only began in 1958. The most common operations performed today were only developed in my lifetime. In the UK more than 300,000 cataract operations are performed each year, almost all as day cases using only local anaesthetic. Artificial hip replacement was invented in the 1960s and each year about 170,000 total hip and knee replacements are carried out. Some 36,000 people each year have cardiac surgery to restore the coronary blood supply or repair and replace damaged heart valves. Every year millions of operations are safely performed that would have been unthinkable 50 years ago.

The risks from the anaesthetic itself are miniscule and life-saving surgery is routinely performed every day on patients in their 80s and 90s. A burst appendix or bowel obstruction is no longer a death sentence. Cancer in the brain can be cut out, aneurysms of blood vessels can be repaired, cataracts can be removed, knees and hips can be replaced. There is nowhere in the body that is out of bounds. Day after day, people live longer, healthier lives as a result.

During the latter half of my career non-invasive techniques began to be developed. Where once it was necessary to perform open surgery many procedures are now carried out non-invasively. Fibre optics and minute cameras on silicon chips have led to a revolution in the ability to use instruments to peer into the bowel, the bladder and every other conceivable body cavity. Many organs can be accessed through blood vessels. Angioplasty is used to improve cardiac blood supply. In an ingenious process a small tube is passed through a blood vessel into the artery supplying the heart. A balloon is then inflated opening up the artery and restoring flow. Abnormal bleeding in the spleen, brain, pelvis and other hard-to-get-at organs can be stopped by blocking blood vessels with injected embolising materials.

Heart valves and major arteries can be replaced with instruments introduced into an artery without the need for open surgery. Surgeons can sit at a computer console directing a robot to perform operations, and they don't even have to be in the same room, or even on the same continent as the patient. Computerised tomography (CT – invented 1972), Magnetic Resonance Imaging (MRI – invented 1977), Positron Emission Tomography (PET – invented 2000) and high quality ultrasound all became available in clinical practice after

I started working. They have transformed the ability of medics to look inside a body and to carry out therapeutic procedures. When I first qualified a doctor wielding a stethoscope tried to make sense of the whooshes and clicks they heard to diagnose a problem with a heart valve. Nowadays, they use Doppler echocardiography to directly view the valves obtaining a clear and accurate diagnosis.

Innovation has paralleled an expansion in our knowledge and understanding of disease, and an enormous increase in the number of effective drugs and other therapeutic options to treat, manage and cure disease. New techniques such as immunotherapy may revolutionise the treatment of cancer and diabetes.

My speciality, intensive care medicine, did not exist when I qualified. The transformation in health care over my lifetime has improved and saved the lives of millions of people. There is every indication that medical innovation is continuing apace allowing all of us to lead longer, healthier and more productive lives.

16 Lifespan: Nutrition

"On the plus side, death is one of the few things that can be done just as easily lying down."
Woody Allen (*Without Feathers*, 1975)

Leaving aside the proven harm of morbid obesity, there is considerable controversy over the role of diet on longevity. Food and drink are one of the great pleasures that make life worth living. How many years of life would you trade if you had to forgo your *steak frites* or *moules marinière* for a starvation diet of lettuce and raw carrot?

Let's at least agree that eating to excess is undesirable. So keep the portions in proportion and don't be greedy.

Dietary advice has probably been dispensed ever since cooking was invented. In *The Cook's Oracle* by Dr William Kitchiner, 1821, the advice at dinner was to consume two wine glasses of sherry, or one of whisky or brandy. After dinner two to four glasses of port or sherry should be taken followed by liqueur as a *bonne bouche* about a quarter of an hour after dinner. A half-hour nap was then counselled followed by another three or four glasses of wine. Add in the copious amount of meat, few vegetables and almost no fruit that was consumed, and it's a wonder that any work got done. The Salisbury steak diet of the 1880s consisted of large quantities of ground beef three times a day with lots of hot water.

Dr James Salisbury, 1823-1905, believed that vegetables produced poisonous substances and therefore were to be avoided. Fad diets are as popular as ever including the Morning Banana Diet, Fruitarianism, the Grapefruit Diet, the Cabbage Soup Diet and the Alkaline Diet.

Most people know what is bad for them. Some chocolate now and again, or modest wine consumption is unlikely to significantly shorten your life, and anyway life is to be lived not just spun out to its greatest extent. It's pretty clear what foods are unhealthy and will not do you any good if consumed in quantities over a long period of time. We are fortunate to be living in a country where cheap, plentiful and nutritious food is readily available. There is also easy access to sweets, chocolates, crisps, fizzy drinks and fast foods. Gratification need not be delayed when there seems to be a coffee shop on every street corner selling muffins and doughnuts. In the same way that a Mars Bar is OK as long as the Coke is diet, and leftover food on a child's plate is calorie free, if the coffee is skinny a brownie can't be fattening. It is no longer sufficient to have a sandwich for lunch. It has to be accompanied by a bag of crisps and a can of drink.

A raw carrot is unacceptable as a mid-morning snack. Even health conscious mothers will put a manufactured "health" drink or smoothie in their child's lunch box. In the popular press we read about people who have to be lifted out of their house with a forklift truck before being loaded into a strengthened ambulance to be taken to hospital. They may be confined to bed and cannot do their own shopping but I guarantee that someone is stopping them starving by supplying a bucket of KFC, a Whopper or two, or a big Mac. Their fridge is unlikely to be filled with lettuce and green vegetables and exercise may well be limited to easing the pressure sores by turning in bed. Not so long ago most people had to walk if they wanted to go somewhere. More labour was manual. Food was bought fresh and was only eaten at mealtimes. Money was tighter and there was little available for non-essentials like snacks.

Nowadays, it takes willpower to ignore temptation and eat well, and determination to exercise in a world where many jobs are sedentary and walking is increasingly a redundant form of transportation.

The potential harm from poor diet coupled with lack of exercise was graphically illustrated in the documentary film *Super Size Me*. For 30 days Morgan Spurlock only ate food from McDonald's. Some might think he overdid it consuming an average of 5,000kcal per day, about twice the recommended intake. At the end of the month he was 11.1kg heavier, had raised blood cholesterol values, a fatty liver, and was experiencing mood swings and sexual dysfunction.

Older people eating more fruit, vegetables and wholegrain bread show better physical performance and less need for social care support. One American study estimated that nearly half the deaths from heart disease, stroke and type 2 diabetes are associated with just 10 dietary factors (see table). Please note that in this study the effect of poor diet was greater in men compared with women, and in blacks and Hispanics compared with whites.

Ten dietary factors associated with deaths

Too much	*Too little*
sodium	nuts and seeds
processed meats	seafood omega-3 fats
unprocessed red meat	vegetables
sugar-sweetened drink	fruit
	whole grains
	polyunsaturated fats

The meal eaten by the healthiest person on the intensive care unit is a large curry with chips and a couple of pints of beer. Or so I told my students. And I wasn't wrong. The on-call doctors would routinely sustain themselves with takeaways before hitting the pub after their shift had finished. Although it may trivialise an important issue it does make the point that humans can live healthily on a wide range of foods. I'm not saying it doesn't matter what we eat. It certainly does make a difference. The amount that we eat can also affect health and this has to be balanced against our needs.

Our food also provides essential ingredients without which we would die of deficiency disorders. In the long run an unhealthy diet will enormously increase the risk of developing a range of diseases.

Sick patients in intensive care are usually unable to eat normally. We feed them by dripping nutrients directly into the blood stream through a vein or introducing food through a tube into their stomach or small intestine. It might not have much flavour but this feed is carefully formulated and typically contains a mix of all the necessary nutrients to sustain life. Over the years various studies have compared different nutritional regimens to see if any one diet has better outcomes than another.

In other words does diet markedly improve survival rates or time to recovery in this group of extremely sick patients?

Various supplements have been investigated including antioxidants such as vitamin C or vitamin E. They have the potential to defend against the free radicals that are inevitable by-products of metabolism, as well as being caused by external factors such as exposure to X-rays, cigarette smoking and air pollution.

Most free radicals are unstable and highly reactive, and cause damage to the lipid in cell walls, to DNA, proteins, carbohydrate and other structures. They are one of the main culprits implicated in ageing through cumulative damage to our cells and have also been blamed for other conditions including clogging of arteries, cancer and eye damage. Another approach has been to boost the body's defence mechanism, the immune system, with immunonutrients including glutamine, arginine and omega-3 fatty acids. An inadequate immune system is clearly undesirable although, perhaps somewhat surprisingly, some diseases are the result of an overactive immune response. Despite many studies looking at nutrition in the critically ill, there is as yet little hard evidence to support the benefit to adding supplements to a standard, complete, balanced diet.

Food is obviously essential to life, although a well-nourished individual can survive for up to about 60 days as long as they have water to drink. When we eat food we take in carbohydrates, fats and proteins as well as a limited list of minerals and essential amino acids. Humans can exist and indeed live long and contented lives on an extraordinary range of diets. The unfortunate panda is doomed to get by on a boring, repetitive menu almost entirely made up of the leaves, stems and shoots of bamboo. Humans are omnivores, which means that we can obtain nutrition and survive on a large range of foodstuffs. Malnutrition is easy to deal with as, by definition, it has an adverse effect. A lack of food or essential dietary nutrients causes health problems and should be avoided. Malnutrition can give rise to vague symptoms such as loss of energy or changes in mood, or to highly specific conditions such as scurvy from vitamin C deficiency.

Gluten is the name given to proteins found in wheat, barley and rye. Coeliac disease is caused by an adverse reaction to gluten and products containing gluten should be avoided. This medical problem can be found in about one per cent of people in Europe and the USA. It has become a fad for others to also avoid gluten with the belief that there may be health benefits. A recent review of the evidence concluded that eating gluten was not associated with an increased risk of cardiac problems. It did conclude that avoiding gluten in the diet may result in a lower consumption of whole grains which themselves are associated with benefits to the health of the heart. Therefore the promotion to non-coeliacs of gluten-free diets to prevent heart disease should not be recommended.

Much of the information about the association between diet and outcomes is from Europe and North America where over-eating is more of a problem than malnutrition. A recent large multi-country study looking at the effect of diet on outcomes found some surprising associations. Not only was high carbohydrate intake associated with an increased risk of death, but so was a low fat diet. Cutting down on fatty food may be a mistake, particularly if carbohydrates are eaten instead. Furthermore, this study did not find that eating fats increased the risk of serious cardiac disease. One commentator made the reasonable comment that the findings just added to the uncertainty about what constitutes a healthy diet.

In the meantime, the bookshelves are full of tomes advocating one diet or another. The Mediterranean diet is often promoted. It is rich in fruit and vegetables, whole grains, fish and poultry. Olive oil replaces butter, little red meat is consumed and a modest amount of red wine is allowed. If this diet is followed along with a non-smoking healthy lifestyle, then unsurprisingly death rates are lower. Other factors are also at work as this benefit may only be seen in the better educated and wealthier. Others diets have their supporters, although some like the Blood Type Diet or Palaeolithic Diet seem pretty far fetched to me. Whatever you choose to believe there can't be much argument for the benefit of a balanced, nutritious diet, avoiding excess and those items thought to be harmful, and it doesn't seem to matter whether the diet is low-fat or low carbohydrate as long as the foods are healthy and nutritious.

In the United Kingdom until recently exhortations were to eat "5 a day" portions of fruit and vegetables, roughly equivalent to 400 grams. The NHS says that examples of a portion of fresh fruit include a 5cm slice of melon, one apple, two satsumas, three apricots, six lychees, seven strawberries or 14 cherries. A vegetable portion is four heaped tablespoons of spring greens, three sticks of celery or a medium tomato. Sadly potatoes don't count no matter how many you eat.

The advice is now to eat 10 portions of fruit and vegetable per day. This is based on research published in 2017.

The authors lumped together 95 published studies and looked at the association between fruit and vegetable intake, and deaths, cancer and cardiovascular disease. The risk of cancer fell as daily intake increased to 600 grams, and deaths and cardiovascular disease fell as intake increased up to 800 grams. The chance of premature deaths decreased by some 10 per cent to 15 per cent for each additional 200 grams a day of fruit and veg although most of the benefit was found with the first 200 to 400 grams. More evidence has emerged to support the benefits of fruit and vegetable although this later study suggested the maximum benefit was found at between 375 to 500 grams per day (4.7 to 6.25 portions of 80 grams per day).

There are all sorts of reasons why fruit and vegetables may be good for you. They contain nutrients, antioxidants, fibre and other potentially beneficially compounds. Although it is clearly a matter of personal choice, higher consumption of fruit and vegetable may be easier if you have personality traits associated with greater intellect, curiosity, social engagement and discipline. Just as plausible as an explanation for the benefits is that the sort of people who eat lots of fruit and vegetables avoid unhealthy processed foods, take lots of exercise, are not obese, don't smoke or drink to excess and have lower intakes of red and processed meat. The study describes an association and cannot say that the fruit and vegetable intake is the cause of the benefits seen.

To get 10 portions in a day you would have to eat something like four heaped tablespoons of spinach, three heaped tablespoons of carrots, a big chunk of cucumber, three heaped tablespoons of butter beans, two broccoli spears, eight cauliflower florets, a small glass of unsweetened fruit juice, half a grapefruit, an apple and two plums. Is there room for much more? It may also be worth finding room to stuff in a few nuts. In yet another study looking at the association between food and undesirable outcomes, nuts have been found to be good for us. Eat five 28 gram servings a week and the risk of heart disease and a heart attack is lowered by 15 per cent to 20 per cent. The biggest benefit is from cashews, pistachios, Brazil nuts and walnuts. Less effect is seen with peanuts. Sadly, peanut butter shows no benefit. Another interpretation is that people with less healthy lifestyles rarely eat nuts but indulge in peanut butter sandwiches.

The body needs 13 vitamins to maintain health. They are vitamins A, C, D, E, K and the eight B vitamins. Apart from vitamin B3 (niacin) and vitamin D, our bodies cannot make these vitamins and they must be obtained through what we eat. Niacin is usually obtained from food although the human body can manufacture it from tryptophan, an essential amino acid.

The main source of vitamin D is from exposure of bare skin to sunlight. Unsurprisingly, deficiency can arise in people who habitually completely cover up when outdoors.

No one disputes that there are diseases caused by deficiencies. Vitamin C (ascorbic acid) will cure scurvy and vitamin B3 (niacin) will work for pellagra (a disease causing diarrhoea, dermatitis, dementia, and death) and there are specific diseases associated with other deficiencies. But with a varied and healthy diet, and plenty of sunlight (for vitamin D) harm from lack of vitamins is unlikely to happen. All sorts of small studies can be found suggesting that vitamins may be beneficial in specific situations. For example, B vitamins may attenuate the adverse epigenetic effects of pollution or prevent Alzheimer's disease-related brain atrophy. Vitamin D can damp down the inflammatory response seen during treatment for tuberculosis.

Because vitamin D is essential for healthy bones it has been extensively studied to see if supplements are worthwhile. A review of all the evidence concluded that supplementation has no important effects on bone density or the incidence of fractures. There is also no clinical trial evidence that increased dietary calcium prevents fractures. Vitamin D supplementation has also been investigated extensively to see if it can prevent colds and flu. Results from individual trials and previous meta analyses have reached conflicting findings. The latest meta analysis gathered together 25 trials with 11,321 participants and concluded that vitamin D was effective at preventing acute respiratory tract infections. Benefit was greatest in those who were very vitamin D deficient. Small studies with limited outcomes, or even meta analysis lumping together many such inadequate trials, are dodgy evidence on which to base recommendations for major lifelong dietary changes.

People who get little exposure to sunlight, have an inadequate diet or malabsorption syndromes are at increased risk of having inadequate vitamin D levels. The current advice is that these individuals should consider taking low dose supplements. Others are unlikely to benefit.

Deficiency in vitamin B3 (niacin) may be a cause of birth defects in women who have an inherited abnormality in the pathway for synthesising the vitamin from tryptophan. In mice engineered to have the same defect, niacin supplementation prevented the malformations. This finding needs further investigation to determine how common it is and as to whether extra niacin should be given to pregnant women.

On the other hand, there are well-described syndromes that can arise from an overdose of vitamins, particularly the fat soluble ones A, D and E. Too much vitamin A can increase the risk of bone fractures. A chronic overdose of vitamin B6 can cause neuropathies. An excess of vitamin C may result in abdominal pain and diarrhoea. Vitamin D supplements taken over a long time have the potential to result in hypercalcaemia (too much calcium) weakening bones and damaging the kidneys. In addition, some supplements, such as herbal weight-loss products, have been shown to contain banned substances that may pose a significant risk to health.

There are also nine amino acids that cannot be synthesised in the body (histidine, isoleucine, leucine, lysine, methionine, phenylalanine, threonine, tryptophan and valine). These are found in a large number of dietary proteins. A strict vegan diet avoids all foods from animals including dairy products and eggs. Without supplements this diet may not provide enough vitamin B12 and other nutrients including calcium and iron. B12 is only naturally found in animal products but is added to some breakfast cereals and soya drinks, and can be obtained from yeast extracts such as Marmite.

Vitamins

Vitamin A: Good sources include cheese, eggs, oily fish, milk and yoghurt and liver. This vitamin is important for the immune system, night vision and some body tissues. An overdose can weaken bones and harm an unborn baby. As large amounts are found in liver the advice is that pregnant women should avoid liver or liver products.

Vitamin B: The eight B vitamins are:
B1 – Thiamin is found in peas, fruit, eggs, wholegrain breads and liver.
B2 – Riboflavin is found in milk, eggs, rice and fortified breakfast cereals.
B3 – Niacin is found in chicken, beef, eggs, venison, many fruits and vegetables, seeds and nuts and fungi. Pellagra can occur if people eat maize as a staple food as it is the only grain low in digestible niacin.
B5 – Pantothenic acid is found in almost all meat and vegetables.
B6 – Pyridoxine is found in a wide variety of foods including pork, poultry, bread, vegetables, peanuts and milk.
B7 – Biotin is found in a wide variety of foods.
B9 – Folic acid is found in many foods including broccoli, spinach and Brussels sprouts. Supplementation is recommended for pregnant mothers, as deficiency is associated with foetal nerve tissue defects such as spina bifida. It also helps form healthy red blood cells and a lack can cause anaemia.
B12 – Cyanocobalamin is found in meat, fish, milk, cheese and eggs. Vegans may become deficient unless they take supplements.

Vitamin C: This is found in citrus fruits, but also a wide range of other fresh fruit and vegetables including potatoes, Brussels sprouts, broccoli, peppers, strawberries and blackcurrants. Many animals (but not humans) synthesise their own vitamin C so it can be found in some meat and fish, particularly if it is not cooked. This is why the Inuit can survive getting their vitamin C from raw caribou liver, kelp and seal brain. It is also why during Shackleton's ill-fated Antarctic expedition in 1915 his men survived a winter on Elephant Island without scurvy by eating raw penguin and seal meat. There is little evidence that vitamin C can prevent or fight colds.

Vitamin D: Ultraviolet-B radiation stimulates vitamin D production from 7-dehydrocholesterol in the skin. Sources also include oily fish, red meat, liver and egg yolks. Some foods have added vitamin D. These include fat spreads and breakfast cereals. Unless fortified with vitamin D cow's milk is not a good source. Vitamin D deficiency causes rickets with bone deformities and pain. Those at risk of rickets include people who rarely go outdoors or who cover up when outside, and the darker your skin the more exposure to sunlight is necessary.

Vitamin E: This is found in many foods including soya, corn and olives as well as nuts and seeds. It is necessary to maintain healthy skin and eyes and the immune system.

Vitamin K: This vitamin is in green leafy vegetables, vegetable oils and cereal grains. Small amounts are in meat and dairy foods. It is essential for blood clotting and for wounds to heal.

Some two-thirds of people living in the United Kingdom take dietary supplements. These include vitamins and minerals as well as other substances such as bee pollen, ginseng, garlic, green tea, omega-3 fatty acids and the plant extract resveratrol. The market for supplements is worth hundreds of millions of pounds a year, and lifestyle gurus, self-help mentors and dietary experts are all at hand to lighten people's wallet in the search for increased vitality, stronger bones or enhanced immunity. In London, adverts adorn the back of buses aimed at healthy people who just happen to be pregnant, male, or female, or have hair, nails or eyes. One leading manufacturer has 10 different products focussing on the brain and 19 if you are more than 50 years old. Apart from vegans who knowingly take a diet associated with deficiencies, there are a few other groups of people who may benefit from supplements. These include pregnant and breast-feeding women, the very young and the very old. As for the rest of us, the benefits of taking supplements have not been confirmed through robust research and indeed in some cases they may be harmful.

The United States General Accounting Office (GAO) has reviewed anti-ageing and alternative medicine products marketed to America's senior citizens (see weblink in References). These included such things as ginkgo biloba, ginseng, St John's wort, glucosamine, fish oil and melatonin. At least 10 per cent of older people in the States regularly take one or more of these supplements. The summary of the report stated "… anti-aging therapies may pose a potential harm to senior citizens.....a variety of frequently used dietary supplements can have serious health consequences for seniors". Observational studies have suggested that diets high in omega-3 (polyunsaturated) fatty acids may protect the brain against dementia and cognitive decline. Sadly more rigorous studies have, so far, failed to show benefit.

The colour supplements and glossy magazines regularly carry articles about so-called superfoods that purport to have health advantages. These claims are usually made on the basis that some constituent has a specific beneficial effect. Garlic, for example, will reduce cholesterol, but you would have to eat some 28 cloves a day to match the doses used in the lab. It is a big leap to go from something that may have a beneficial effect to proving that it actually makes a difference when eaten as part of a diet.

In the laboratory antioxidants can undoubtedly protect against the harmful effects of free radicals. However, a review of 78 randomised studies with 297,707 participants found no evidence that dietary antioxidant supplementation did any good. On the contrary, beta-carotene and vitamin E may actually increase mortality, as may higher doses of vitamin A. It is entirely possible that removing free radicals from the body interferes with the body's defence mechanisms. There are few scientific studies to support most of the claims relating to superfoods.

Where there is good quality research the conclusion is usually that there is no benefit, or at best one that is modest. Besides, anyone who thinks that eating a few superfoods will negate the effects of a big Mac and fries is kidding themselves.

Superfoods

Blueberries: Low in calories, high in nutrients including antioxidants.
Claimed benefits: Protects against heart disease, some cancers and improves memory.
Evidence: May decrease risk of heart attack, in a large observational study. Is this association or causality? In a small study, decreased blood pressure. No evidence to support benefits for cancer or memory.

Goji berries: Contains vitamin C, B2, A and iron, selenium and other antioxidants.
Claimed benefits: Boosts immune system and brain activity, protects against cancer and heart disease, improves life expectancy.
Evidence: No reliable evidence to support these claims. Scientific studies used purified extracts at much higher concentrations than there are in the berries.

Chocolate: Cocoa contains iron, magnesium, phosphorus, zinc and antioxidants.
Claimed benefits: Reduces blood pressure and thus helps prevent heart disease and strokes, prevents cancer, relieves stress and helps leg ulcers heal.
Evidence: Dark chocolate may reduce blood pressure by a small amount (2-3 mm Hg) in the short term. Other benefits are weak or unproven. The studies focus on cocoa extracts, not chocolate. Chocolate as normally consumed is stuffed full of sugar and fat.

Oily fish: Contains vitamin D, some B vitamins, omega-3-fatty acids and antioxidants.
Claimed benefits: Prevention of cardiovascular disease, prostate cancer, dementia.
Evidence: There is strong support for the view that eating oily fish may decrease the risk of heart disease. There is some evidence to suggest that it may reduce the chances of getting age-related macular degeneration in the eye but further research does not show any benefits in slowing the progression of this disease in those who had already developed it.
There is no evidence that it can reduce the risk of developing dementia. In healthy mice tests have shown that it does not extend longevity.

Wheatgrass: Contains chlorophyll, vitamins A, C and E, iron, calcium and magnesium.
Claimed benefits: Protects against inflammation, builds red blood cells, improves circulation.
Evidence: No sound scientific evidence of benefit.

Pomegranate juice: Contains fibre, vitamins A, C and E, iron and other antioxidants.
Claimed benefits: Prevents heart disease, lowers blood pressure and decreases inflammation and cancer risk.
Evidence: It may prevent osteoporosis, although the evidence is from a mouse study. It may also slow down the progression of prostate cancer. Some small studies show that it may improve arterial blood flow.

Green tea: Contains B vitamins, folic acid, manganese, potassium, magnesium and antioxidants.
Claimed benefits: Boosts weight loss, reduces cholesterol levels, combats heart disease, prevents cancer and Alzheimer's disease.
Evidence: There is no evidence of benefit against cancer. It may lead to a small reduction in cholesterol and a small fall in blood pressure.

Broccoli: Contains vitamins C, A and K, folic acid, calcium, fibre and antioxidants.
Claimed benefits: Can fight cancer, reduce high blood pressure, heart disease and the incidence of diabetes.
Evidence: May reduce the risk of mouth, throat and stomach cancer.

Garlic: Contains vitamins C and B6, manganese, selenium and antioxidants.
Claimed benefits: May be effective against high blood pressure, heart disease, high cholesterol, colds and some cancers.
Evidence: May help lower blood pressure and cholesterol. Doubtful, if any, benefit for cancers.

Beetroot: Contains iron and folic acid and antioxidants.
Claimed benefits: May lower blood pressure, boost exercise performance and prevent dementia.
Evidence: May cause a small fall in blood pressure and some improvement in exercise performance but not in elite athletes.

17 Extending life: Introduction

"You only live once, but if you do it right, once is enough."
Mae West, (actress, 1893-1980)

Ageing is increasingly seen as a disease. Once it is accepted as such then it makes sense to try and find a cure. Money may not be able to buy love, and it certainly can't buy long life, at least not yet. For some wealthy individuals wishing to live a bit longer this is an important area for research, and they are willing to pay handsomely for the opportunity to benefit. As preventing ageing moves into the mainstream, companies see that investment now may produce healthy returns in the future. A quick Internet search will turn up organisations with names including the Maximum Life Foundation, International Longevity Alliance and BioViva, all trying to find a way to extend the human lifespan.

Until now the main focus of medical research has been on the causes and treatment of individual diseases. There are institutes dedicated to research on cancer, heart, lung, blood problems, infectious diseases, diabetes and so on. The biology of ageing encompasses all these organs and more. Many age-related diseases appear to share common metabolic causes. If so there may be something fundamental in the process of getting old that can be targeted and will increase healthy lifespan. If it can be done it is likely to be a more effective strategy than treating individual illnesses. Genome wide association studies have tried to identify genes that are common to age-related diseases. There is some evidence to show that genes, such as the one involved in cholesterol metabolism, (APOE – apolipoprotein E genes) are strikingly enriched in several age-related diseases. Apolipoproteins bind lipids so that they can be transported through the blood and lymphatic system and were identified as the important link in this process. If the same mechanisms that cause age-related disease also directly cause ageing, then the expectation would be that treatment would deliver longer healthier life.

There are several ways in which it may be possible to extend life significantly beyond the 100 to 120 years that are achievable today. At the moment they are all in the realm of science fiction, although some seem considerably more unlikely than others. No one can tell you what may work because no one knows. We do, however, have some sense of the direction of travel and I will try and indicate where developments may take us.

One of the challenges is to identify the mechanisms of ageing. Given the number of years people live, it is difficult to prove that an intervention in humans is effective. It is possible that accurate biomarkers will be useful surrogates when assessing the effectiveness of interventions. Telomere length and the pattern of DNA methylation are both candidates for this. Human studies are already underway with interventions specifically aimed at modifying ageing. Animal models will always be one of the main ways that interventions are tested. Companion animals such as cats and dogs have attractions as they share the same human environment and ageing pathology, although have a different diet, lifestyle and (in our household at least) bed to lie in.

Ageing is global, affecting all tissues and all cells. Much can be done to manage its ravaging effects. I have already alluded to the beneficial effects of diet, exercise, mental stimulation and social interaction. To extend lifespan meaningfully beyond what is currently possible, medical science is going to have to up its game.

There are some reasons to be optimistic. To start with considerable progress has been made in elucidating some of the ageing mechanisms. The genetic code has now been cracked and can be read and written. Not only do genes contribute to ageing but, in theory, the potential exists to modify them in unique ways to delay the ageing process. We also know that somatic cells, the normal specialised cells of the body, can be turned back into stem cells. This is a form of rebirth, a resetting of the mechanism back to zero from where it can start again. Although not essential, it is unlikely that much progress will be made without a better understanding of how ageing happens. The chances of stumbling across the elixir of life are improbable so any progress is likely to be the normal combination of hard slog with the occasional flashes of inspiration.

Research into ageing is now mainstream. The NIH (National

Institutes for Health) in the United States is the largest biomedical research agency in the world. One of the subgroups of the NIH is the NIA, the National Institute on Aging with an annual budget of some $1.6bn. It funds research into Alzheimer's, understanding the biology of ageing, and interventions with the potential to extend lifespan and delay disease and dysfunction.

The application of knowledge coupled with a willingness to improve the lives of fellow citizens has been very successful at allowing most people to achieve close to their potential lifespan. More recently there is evidence to suggest that what is achievable is getting longer. More people are living into their 90s, 100s and beyond, something that was extremely rare in earlier generations. This has been achieved by small incremental changes, not by startling new insights or innovations. Attention to detail and doing to the best of our ability what we know to work can achieve remarkable results.

The American writer and entrepreneur, Jim Rohn (1930-2009) said: "Success is a few simple disciplines practiced every day; while failure is simply a few errors in judgment repeated every day."

There is no doubt that lifespan can be experimentally manipulated. Many genetic defects and environmental challenges will shorten lifespan. There are also genetic and environmental interventions that have been shown to extend healthy life in creatures such as the nematode worm, fruit fly and laboratory mouse. Extending lifespan is subtly different to delaying ageing, as interventions may prevent common causes of death such as infection or cancer rather than altering ageing itself. There is, however, some cautious optimism to believe that ageing itself can be slowed, temporarily arrested or even reversed. The roundworm *C. elegans* (*Caenorhabditis elegans*) normally lives for about two weeks. However, in some circumstances a metabolically inactive state can be induced prolonging survival for months. There are many other examples in nature where development is arrested for prolonged periods. This is common in seeds and bacterial spores demonstrating that biological ageing can be uncoupled from the passage of time.

Although resetting the biological clock seems far-fetched it is achieved with every human foetus. Here a sperm and egg, each of whose age may be measured in decades, fuse together and rewind the chronological clock to start anew. It has also been shown that the DNA in the nuclei of adult cells can give rise to viable embryos without any apparent signs of premature ageing. This is the process that created Dolly the sheep.

As previously mentioned, is possible to create pluripotent stem cells from differentiated adult cells. In other words to turn mature established cells that know what job they have to do back into youngsters with the potential to be any cell they care to choose. If that is not rejuvenation I don't know what is. If the blood of a young animal is circulated through the blood vessels of an old animal there are rejuvenating changes in the old animal. In contrast the young cells start to look old. Furthermore, gene regulation has been altered in mice resulting in less skin ageing. Taken together with the ability of rapamycin to reverse markers of old age there is optimism that ageing can be interfered with.

Although it may be possible to prolong human life there are relevant questions about whether we should. These techniques go well beyond treating human disease but are aimed at overcoming the limits of human nature. Objections have been raised on several levels.

Will developments lead to two different types of being, those doomed to be trapped within their limited bodies and others who have access to the treatments that will enhance their powers and prolong their lives?

Discrimination will be based not on superficial irrelevant qualities such as skin colour, but on measureable outcomes. In the future the rich may not only be richer but may be the only ones who can afford to pay to be able to live much longer, healthier lives. Barriers of wealth and power will limit individual choice and autonomy. Some critics see artificial enhancement of the human condition as a threat to morality because the essence of human nature is under attack. Trans species biology and neuroengineering may become reality. What sort of creature is a human with animal or artificial genes? At what stage does a human/computer cyborg become more machine than man? Changing the genome alters what we are as humans. Some fear it may have unintended consequences for individuals who will breed damaged offspring. It may spread harmful genes into the biosphere contaminating living organisms, including food, dooming humanity and life on earth. Overpopulation, already a serious concern, becomes even more likely with few people dying. Competition between generations might arise and scenarios from war to ossification of society can be envisaged.

The current revolution in biology may be a slippery slope that will lead mankind to conflict, disorder and destruction. The innovations I describe, and those that may come, have the potential to deliver enormous benefits to many. In common with much progress it also has risks and raises legitimate questions that society will have to address. I have not set out to review in detail the ethical and moral issues that may arise. I do touch briefly on some of the implications when imagining a world of millenarians in the final chapter.

I am, however, once again reminded of the early days of anaesthesia. Before the first practical demonstration of ether anaesthesia in 1846 it had not been possible reliably to relieve pain and suffering in patients undergoing surgery. From our perspective, today, we might think there was universal acceptance of this new discovery. After all what can be wrong with painless surgery? But at that time there were plenty of people decrying the meddling with God's creation and ordering of things. Who are we to interfere? There were others emphasising the importance of pain as a diagnostic and healing sensation, to be balanced against the undoubtedly harmful possibilities of receiving an anaesthetic. Indeed the City of Zurich outlawed anaesthesia altogether stating "Pain is a natural and intended curse of the primal sin. Any attempt to do away with it must be wrong." Other clergy quoted the Bible that states "In sorrow thou shalt bring forth children" clearly railing against pain relief in childbirth. Others saw anaesthesia as a decoy from Satan.

Some eminent doctors spoke out against anaesthesia on both moral and medical grounds saying pain was essential to life. The American Temperance movement regarded the use of ether as a form of intoxication particularly posing a threat to the virtue of female patients. Some of these objections died away after Queen Victoria took chloroform during the delivery of her eighth child, Prince Leopold. Perhaps anaesthesia was the first step on a slippery slope. Taking away pain and suffering during surgery interferes with the world God created for us. If we can do this then should we meekly accept the status or rank into which we were born? Should we tolerate the pain and suffering inherent in child labour or exploitation of the poor, women or other minorities?

Medically, once we accept anaesthesia, there is no logical place to stop along the path to organ transplantation, stem cell therapy or genetic manipulation.

18 Extending life: Evolution

"If all else fails, immortality can always be assured by spectacular error."
John Kenneth Galbraith (Canadian-born US economist, 1908-2006; *Money, Whence It came, Where It Went*)

Popular mythology has us believe that female black widow spiders kill and consume their mates after insemination. We can speculate from an evolutionary point of view why this might happen. The female gets a good meal without having to make much effort, the spider equivalent of a Domino's pizza I guess. It also gets rid of undesirable fathers who loaf around the place messing up the web and generally not contributing to bringing up junior. As for the male it might buy him more time to get on with the business of passing on his genes and some even leave their detachable pulp, a spider's penis, inside the female blocking access for other males. There's little doubt that this behaviour goes on, and the male can sometimes be observed throwing himself into the jaws of his mate. However, it's not quite as straightforward as all that. Many male spiders don't want to be eaten and try to avoid this gruesome fate, some females tuck in before sex has begun and it's not unknown for males to gobble up the females of their species. As we're unable to question spiders about their motivations any theories about why this happens have got to be highly speculative.

Semelparous species reproduce once and then die (see page 78). As humans we're fortunate that we get to have sex more than once. We are also allowed to hang around to see our offspring grow up, to be a credit or a disappointment to us. For the unfortunate squid, octopus and salmon death is clearly genetically controlled.

This digression is meant as an introduction to natural selection. This is the key mechanism in evolution. What it means is that individuals with desirable attributes survive and have more opportunity to reproduce and pass on their characteristics. Over many generations this reproductive advantage leads to more offspring possessing the desirable traits. Evolution through natural selection is the mechanism by which individual species develop. As most people know, Charles

Darwin elaborated this theory after visiting the Galapagos Islands and other locations during his five-year journey on HMS Beagle. He was intrigued by the slight variations in tortoise shells and finch beaks between different islands. This culminated with the publication in 1859 of his seminal book *On the Origin of Species*. No longer did a divine being have to be invoked to explain how animals changed, adapted and evolved. Darwin's book elegantly laid out the mechanism to explain how this happened.

Selective breeding of plants and animals has been going on since the dawn of civilisation. This has produced new varieties of cereal, breed of dog, or type of horse. Human intervention over many hundreds of years has developed a wolf into dogs ranging from Chihuahuas to Afghans. Selective breeding by mankind leads to animals and plants that have the characteristics the breeder desires. Pressure from the environment does the job of selecting organisms that have superior survival and reproductive traits in that environment compared with others. Natural selection depends upon the environment and survival does not imply that the creature in question is more developed or superior to others that may have gone before. It is just better adapted to its current circumstances. Darwin described how evolution worked but at that time he did not know the mechanism by which that could take place.

The Augustinian friar Gregor Johann Mendel provided clues to that question. He experimented with pea plants and showed how certain characteristics such as seed shape, flower colour and plant height could be transmitted. However, the significance of his work was not widely recognised for another 30 years until the end of the 1800s. The key understanding is that characteristics are not a "blend" of the mother's and father's characteristics, but that the individual units of inheritance, the genes, are passed on intact from one or other of the parents.

The theory of evolution has always had its fierce critics, none more vocal than in the United States. Creationists have tried with intermittent success to prevent the theory being taught in schools. In 1982 Judge William R Overton overturned a law in Arkansas writing: "While anyone is free to approach a scientific inquiry in any fashion they choose, they cannot properly describe the method as scientific if they start with a conclusion and refuse to change it regardless of the evidence…". For us to accept the theory of evolution we must believe that the Earth has been around for many billions of years, that organisms can change and adapt without any external interference, and that mankind is an animal related to and descended from apes, one of many species that have evolved on our planet. For some people it has been impossible to believe that man is not created and is not, in some way, the pinnacle of creation. Even among rational scientists there were competing theories and explanations of evolution until the 1930s.

Over the years the theory of natural selection has been developed and refined. However, the basic concept remains unaltered. With the understanding of genetics the mechanisms by which evolution takes place have now been revealed.

Let us consider what this means in terms of human lifespan. Humans breed relatively slowly and normally have few children. Each child is dependent for many years and takes a long time to achieve maturity and independence. The father may want to go off and spread his seed around but there is probably an evolutionary advantage if he hangs around, or at least is made by the Child Support Agency to contribute to the child's upkeep. As an aside it's generally easy to identify a child's mother. However, studies suggest that up to 30 per cent of children are fathered by someone other than the man who believes he is the biological dad.

It's probably also a good evolutionary idea to have grandparents at hand to help with the childcare. The younger, fitter and more active parents can go out and earn a living or enjoy their social life, while the more sedentary older folk can fetch and carry from school, do the ironing and help with the homework. A little bit of wisdom may also be imparted along with a couple of jokes and boring anecdotes about what it was like in the war. After all, the grandparents have a genetic stake in the survival and wellbeing of their grandchildren. As time goes on this genetic equation works against the oldies. There's no point in wasting resources on people who can't make a contribution and need looking after themselves.

Humans are already extremely long-lived compared with their closest relatives, chimpanzees and bonobos. Whereas well-cared for individuals of the species *H. sapiens* may reasonably expect to make it into their 80s, a chimp, even in captivity, will be unlikely to make it beyond 60 and bonobos to about 40 years of age. There is no mechanism for evolution to select very long-lived humans. They've had their kids, helped raise their grandchildren and now is their time to make way for the younger generations. Obviously we'd rather marry someone from a family who aged well, but the thought of having ancient decrepit grannies that have to be accommodated in what used to be the living room is not an inducement to romance. I could even speculate that the evolutionary pressures are now for grandparents to die off early so their children can inherit, as the only way that they will ever be able to afford a house of their own.

In fact, modern interventional medicine may weaken this drive to selection. Individuals with genetically undesirable traits are kept alive long enough to breed, the state provides support when feckless fathers and mothers fail their children, and *in vitro* fertilization removes the need for sperm to compete by choosing individual ovum and sperm to produce a baby.

The Darwin Awards are given to the individual who has contributed to human evolution by selecting himself or herself out of the gene pool. This somewhat macabre trophy can only be awarded posthumously. It commemorates those who have eliminated themselves in an extraordinarily stupid manner from being able to pass on their genes. There are five criteria that have to be achieved:

- **Being dead or rendered sterile thus unable to pass on genes**
- **Judgment that is astoundingly stupid but not commonplace (smoking in bed would not count as this happens far too often)**
- **Be the cause of one's own downfall**
- **Being free of mental defect, i.e. they should have known better**
- **The event must be reliably documented from reputable sources**

Award nominees include a 19-year-old factory worker who decided to climb into the tiger enclosure at Delhi Zoo, the gullible acolyte who volunteered to be killed so that a trainee holy man could resurrect him, the man who lay undiscovered for five years in an underground vault having been electrocuted stealing copper wire, people jumping off rocks, standing in front of trains, … and so on.

Not many individuals are taken out of the gene pool by their own stupidity so this mechanism cannot be relied upon. In short it is unlikely that evolutionary pressures will select for extreme longevity.

If the gene is to play a part in extending our lifespans we will have to intervene directly with the genome (see chapter 23).

19 Extending life: Reaching for immortality

"The only thing wrong with immortality is that it tends to go on forever."
Herb Caen (*Herb Caen's San Francisco: 1976-1991*)

It was Benjamin Franklin who wrote "In this world nothing can be said to be certain, except death and taxes." Some large international businesses seem to have found a way of avoiding or, at the very least, minimising the amount of tax they have to pay. Similarly, there are some people who believe that death can be avoided, or at least delayed for a very long time.

Religions have always had a lot to say about eternal life, or rather what happens after our time on earth ends and we move onto whatever their particular view is about what happens next. Whatever anyone believes, hopes or prays for, most of us are in no hurry to shuffle off our mortal coil and move onto the next phase. So while we're alive we might as well make the most of it.

Immortality, or even better, eternal youth, has long been a staple of myths and legends. Methuselah, grandfather of Noah, is said to be the oldest person in history. According to the Bible (Genesis 5:21-27) he lived for 969 years becoming a father at the age of 187. Alternatively it could be a mistranslation and he was really only 78 or a more creditable 97, or perhaps he never existed at all.

The Holy Grail is the cup from which Jesus drank at the Last Supper. Legend has it that King Arthur and his knights searched for it. The Monty Python team made a film about it and Indiana Jones went chasing after his father who had gone missing while on its track. In some stories Sir Galahad is the knight who finds the Grail and achieves immortality. In Japanese mythology eating the meat of a Ningyo, a sea creature that is an omen of bad luck, brings the curse of eternal life.

The Greek Gods did not themselves die but they could punish mortals by conferring immortality. Thus King Sisyphus was made to roll a huge boulder up a hill only to have it roll back down. He was forever condemned to carry out this pointless task. In the Chinese epic *Journey to the West* Sun Wukong, the Monkey King, was bestowed with 1,000 years of life by eating a Peach of Immortality. Aldous Huxley wrote about a rich man's quest for immortality. In his book *After Many a Summer,* the curse is man's continued evolution turning from a human foetal ape into a grotesque simian adult. In Mary Shelley's novel, *Frankenstein: or, The Modern Prometheus,* Victor Frankenstein discovers the secret of life itself and uses it to create a creature out of old body parts.

In modern culture, Dr Who lives forever by metamorphosing whenever it's time for a new actor, and several characters in the Harry Potter books are immortal. It is also the basis for many movies including Highlander starring Sean Connery, and various versions of Dracula, although he is not so much immortal as undead, as are the countless vampires and zombies who have appeared on the silver screen. Many science fiction stories explore the issues of eternal life. Throughout all these examples immortality is commonly a punishment or curse, and not a desirable goal. It always seems to come with disadvantages and drawbacks.

Despite the doubtful benefits conferred by immortality in the literature, many attempts have been made to find the elixir of eternal life. In ancient China various emperors are reputed to have searched for it. One example of where doctors really got it wrong relates to the Jiajing emperor in the Ming Dynasty. The particular elixir mixed up by his alchemists contained mercury and, when ingested, the poison killed him. Amrita, the nectar of immortality, has been described in Hindu literature, and in Europe, St Germain was reported to be several hundred-years-old from imbibing the elusive elixir.

During the entire 20th century some 5.5 billion people died and each year a further 56 million come to the end of their lives. Enormous strides have been made enabling many people to remain active and healthy into ages that would have seemed extraordinary in previous generations. It no longer seems unusual that an 80-year-old has climbed Mount Everest, a 75-year-old participates in single-handed ocean racing or that a centenarian has run a marathon. Although death has not yet been banished it has been pushed back many decades.

In the meantime, medical knowledge and innovation is expanding exponentially. Try to imagine a world where almost no one dies and everyone can reasonably expect to live to be 1,000. Up to this moment everyone who has ever lived has expected to die. Outside of science fiction and theology, life eternal is probably a step too far. I'm going to lower my sights and consider whether 1,000 years is possible, how it might be achieved and speculate a little about what it might mean.

20 Extending life: Maintenance and repair

"Millions long for immortality who don't know what to do with themselves on a rainy Sunday afternoon."
Susan Ertz (*Anger in the Sky*, 1943)

At present anti-ageing management is based on a piecemeal approach of fixing what is broken. This ranges from traditional tissue and organ repair, transplantation and replacement, as well as targeted treatments with stem cells. These treatments are often life saving in the short-term but are essentially transient, short-term and limited interventions with little or no impact on substantially extending duration of life. Nonetheless incremental improvements in conventional medical care will continue to deliver better ways of treating disability and disease.

Non-medical technology is already helping people live longer, safer, healthier and more independent lives. A mobile phone allows the elderly to keep in touch and call for help in times of crisis. For those with access to the Internet and the ability to use it, groceries and other supplies can easily be ordered and delivered to homes. If they can't order online, relatives or friends can do it for them. Computerised home assistants respond to commands, carry out tasks, answer questions and provide entertainment. Many more aids are available to help the old and infirm be safe in their own homes, and travel has gradually become easier for those with mobility problems. Within a few years self-driving cars will give freedom of movement to those who can no longer drive. Hearing aids and speakerphones can compensate for the common loss of hearing that comes on with age.

Smart health monitors are starting to become available. This is wearable technology that continuously monitors physiological variables including heart rate, skin temperature and blood oxygen values. Physical activity can be tracked, often in conjunction with GPS location, and sleep can be monitored. These devices are popular and as of July 2015 there were more than 500 different health care-related wearable products on the market with more than 34 million devices sold. How often do you check your temperature or oxygen saturation or even heart rate? By contrast some of these devices allow up to 250,000 daily measurements to be taken from one individual. They have been shown to give early warning of the onset of infection and to identify fatigue. Portable biosensors are the first step in continuous physiological monitoring that will almost certainly play an important role in monitoring health status in the future, allowing early warning and aiding diagnosis. Biomarkers in the blood, including the DNA of circulating cells, have been shown to pick up recurrent cancer much earlier than would otherwise be detected. A Google (Alphabet) company called Verily Life Sciences is developing ways of continuously collecting health data to better understand ways to predict and prevent disease. Projects include continuous blood glucose monitors, wireless sensing eye lenses, bioelectronic medicines and surgical robots. It may even be possible to replace lost senses such as vision and hearing with a vibratory vest. Sounds crazy? Check it out at *www.eagleman.com/research/sensory-substitution*. Progress will undoubtedly be made on a wide range of technological innovations, of which some may well be unanticipated and life changing.

Advances in technology brought us the CT scan, the MRI scan and PET scanning. Now artificial intelligence software is helping make use of the data provided by the MRI. By collecting data from 30,000 points each time the heart contracts a 3D model can be constructed of it. Pulmonary hypertension is a life-threatening condition of high pressure in the blood vessels connected to and within the lungs. Computer analysis of heart scans in patients with this condition is much more accurate than doctors at predicting survival at one year. Intelligent analysis and expert support systems will help doctors make the most of clinical information and guide them towards the most appropriate treatments.

Scientists are in the early stages of integrating technology with living organisms. A part-human, part-machine organism is a cyborg. Humans already walk around being kept alive with artificial heart pacemakers and defibrillators. Brain stimulators are used to treat patients with Parkinson's disease and epilepsy. Cochlear implants to restore hearing have been called bionic ears. Tiny electrical implants are being developed to read and modify electrical signals that pass along nerves. The hope is that modifying these biological signals will help treat disorders as diverse as inflammatory bowel disease, arthritis, asthma, high blood pressure and diabetes. Bioelectronics or electroceuticals, as this new branch of medicine is called, is being supported by major pharmaceutical companies including GSK. Lens implants have solved the problem of cataracts for many millions of people. It is now feasible to restore a degree of vision by inserting a microchip within the eyeball to stimulate visual nerves in response to light. An artificial larynx provides a voice for people, such as the late Professor Stephen Hawking, who cannot speak. Joint replacements have successfully replaced worn out hips, knees, shoulders and ankles in millions, relieving pain and restoring function. Artificial hearts are keeping people alive for months and sometimes years. Fully robotic limbs have been built allowing a complex range of motions to be controlled through nerve-muscle grafts. Artificial limbs are already good enough to allow disabled athletes to compete with the able bodied.

In the near future prosthetics may take performance beyond what is humanly possible.

Powered exoskeletons are helping the paralysed walk again. They have the potential to help the weak and infirm regain lost abilities, and will also help healthy people perform well beyond what is achievable with the normal human body.

"Grinders" are people who seek to improve their own bodies with do it yourself artificial cybernetic devices. Transhumanism is a movement that refuses to accept the limitations inherent in our biological bodies. It seeks to overcome biological frailties, including disease and death. Transhumans aim to use biotechnology to transcend our present limitations. This is programmed evolution aimed at improving intelligence, psychological resilience and strength.

Cosmetic techniques such as plastic surgery and orthodontics are already used to produce superficial changes in human appearance. Drugs, such as anabolic steroids and growth hormone, are taken by some individuals to artificially increase muscle mass and density. Nootropic pharmaceuticals, about which more detail is given later in the book, are designed to enhance cognitive abilities.

Gene therapy is already being used to treat disease. The same techniques can be used for genetic enhancement, the insertion of additional genes to produce a change in characteristics. This is one step away from true designer babies where behavioural and physical traits would be pre-programmed into an embryo. If that is a scary thought we are on the threshold of using technology to enhance the human body. Respirocytes are hypothetical artificial red blood cells theoretically able to carry 236 times more oxygen to tissues than an equal volume of natural red blood cells.

Deep brain stimulation is already being used to treat Parkinson's disease, dystonia and tremors. Implantable integrated circuits are being investigated to restore cognitive function in people with brain loss. Implants in the prefrontal cortex of monkeys measurably increase decision-making abilities. There is a real possibility that implantable devices could improve perception, thinking, motor control and even mood. Research is taking place into making brain-to-computer interface a reality allowing the brain to directly control a computer. One vision integrates the human and computer mind for enhanced memory, recall and thinking. In the future, retinal and brain implants may be even better than natural organs so that an upgrade would be advantageous.

The emphasis in surgery is shifting from simply cutting out damaged organs and structures, to maintaining and restoring function. Even today large number of replacement body parts are being inserted. The trend is away from metal, ceramics and plastic towards tissue engineering, where artificial materials and biology combine to produce a more organic solution. Artificial materials have been inserted into humans for many years. Some sutures or stitches used to repair tissues are made of catgut. This is nothing to do with cats, but is made from purified collagen taken from cattle, sheep or goats. These days artificial absorbable polymers such as polyglycolic acid and polylactic acid have largely replaced catgut. Non-absorbable sutures are made from polypropylene, polyester or nylon. Stainless steel wires may be used in some special applications such as closing the breastbone after it has been split to perform cardiac surgery. Synthetic material such as Dacron and ePTFE have long been used to make artificial blood vessels to repair or replace damaged arteries.

Mesh made from polyglycolic acid or polypropolene is inserted to make hernia repairs stronger. Graphene was only isolated in 2004 and is an ultra thin, two-dimensional lattice of carbon. Its mechanical and electrical properties mean that it is being looked at within the body for repairs to nerve, bone, cartilage, muscle and skin. It is also being developed for drug delivery and ultrasensitive biosensors.

Tissue engineering often involves using a manufactured "scaffold" to provide a framework into which cells can grow. This is used to repair blood vessels, rebuild bladders and replace cartilage in joints and elsewhere. Developments have progressed from simple structures up to complex artificial organs such as the trachea (windpipe) using pig organs to rebuild the trachea and oesophagus in children. The pig organs are stripped of cells and seeded with the recipient's stem cells. Materials include polyester and organically based substances such as collagen or fibrin. Scaffolds can be made in several different ways including 3D printing.

Stripping an animal organ of its cells takes away the risk of an adverse immune response. This has been tried with monkey kidneys onto which human embryonic stem cells were seeded. Ways have been found of incorporating organic growth within the biomaterials. Blood vessels can be encouraged to proliferate with chemicals such as vascular endothelial growth factor.

Stem cells can be implanted encouraging tissue to grow and bone to accumulate.

Over time the biomaterial becomes an integral part of the human body and the non-organic components may even dissolve away leaving a completely organic structure. The hope is that damaged or destroyed tissue can be repaired and replaced, including bone, the gut and airways. I anticipate seeing more implants with biological and artificial combinations to repair tissues and rebuild failing organs.

A "nanomedibot" is a device a few molecules in size that is able to perform a pre-programmed medical task. The idea is to inject them into the bloodstream to repair damaged or diseased tissues. For example, they could be designed to recognise, seek out and destroy cancer cells. If they are designed to detect chemicals and toxins these minute robots could be used to give early warning of organ failure and to monitor an individual's health. When reality catches up with these predictions then, in the not too distant future, our arteries and veins will be carrying millions of minute, constantly vigilant, workers ready to detect and treat a range of medical conditions.

In chapter five, I discussed what is understood about how memories are stored. Amazingly, it may be possible to alter memories or even implant learned skills. Electrical hippocampal stimulation can pass on learned knowledge from trained to untrained rats. This steps firmly into the realms of science fiction with the prospect of directly transferring information from one brain to another. Imagine learning juggling or French simply by having a brain implant with patterns taken from an expert. In other experiments four rat brains were connected together using implanted electrodes, and simple brain-to-brain communication in humans has been demonstrated.

The billionaire Elon Musk has set up a company called Neuralink to explore connecting human brains with computers to download and upload thoughts, to create a merging of human and machine. A company called Kernel has the stated objective of working to maintain brain vitality and unlock its trapped potential. Already brain-controlled muscle stimulation has been used to restore partial arm control in a quadriplegic person.

Experiments have been done to directly link a human brain over the Internet to remote devices. This is the first step to controlling a robot or mechanical avatar, able to live an individual's life remotely under their control.

It is even possible to buy a kit called a RoboRoach to control the left/right movements of a cockroach. As I type they are available – cockroach not included – from Amazon USA (not the UK) for only $119.99 with free two day shipping. Cyborg tissue has been created by embedding a three dimensional network of biocompatible, nanoscale wires into human tissue. The developers said that it was "about merging tissue with electronics in a way that it becomes difficult to determine where the tissue ends and the electronics begin".

In the future, technology will play a growing role in making life easier, preserving function and replacing worn-out tissues and organs.

21 Extending life: Calorie or dietary restriction

"Every man's life ends the same way. It is only the details of how he lived and how he died that distinguish one man from another."
Ernest Hemingway, (foreword to AE Hotchner's biography, *Papa Hemingway*, 1966)

Calorie restriction (CR) deserves a special mention as one possible way of extending lifespan. This effect in rats was first described in 1935 so it has a long history. CR is emphatically not the same as starvation but is a reduction of energy intake without malnutrition. Many different creatures have been subject to CR ranging from yeasts, spiders, silkworms and fish, up to mammals including dogs and primates. In one example, mice fed a nutrient rich but calorie-reduced diet lived for 53 months compared with 35 months in controls. Most of these studies have shown that CR dramatically delays the onset of the signs of ageing, and increases the length of life, and it is a consistent finding across many species and many studies. At the physiological level there is an early acute phase followed by several weeks of adaptation until a new stable state is reached. A lower body temperature, lower glucose and insulin values, and reduced body fat and weight characterise this. In evolutionary terms, this may represent an adaptation to periods of scarcity where the ability to slow ageing and delay reproduction would give a competitive advantage.

Dietary restriction may be a better term than CR. When *Drosophila* the fruit fly is overfed with essential amino acids it has a shorter lifespan. Perhaps an imbalance of dietary amino acids is important rather than calorie intake alone. A word of caution is necessary. Rodents and insects appear to respond differently to CR. In some studies a decrease in caloric intake lessens rather than prolongs the lifespan of flies.

There have so far been three independent studies using rhesus monkeys to investigate the effects of CR on longevity. This is likely to provide the best information about the effect of CR in humans. Monkeys are long-lived as well as difficult and expensive to keep for long periods in captivity. There won't be many more of these studies performed in the near future. The first of these studies only had eight male monkeys on a CR diet and provided limited information. The other two studies specifically focused on the impact of CR in healthy animals. One study reported that CR improved health and age-related survival. The other failed to show a statistically significant benefit from CR. As with much medical research this raised as many questions as it answered.

An in-depth analysis has compared both of these studies. There were differences in the protocols that may be relevant. These include the age of the animals when they were enrolled in the study, the age at which CR was started, the pattern and amount of feeding, and whether the monkeys were Indian or Chinese. When all these factors were taken into account the overall conclusion was that CR did increase survival, and that it worked even when the animals were quite old before the diet was imposed on them. There are a few caveats to this. First, there may be a gender difference as in one of the studies female monkeys being calorie restricted didn't live longer but the males did. Second, CR may work when implemented in adults but may be of no benefit when started earlier in life. Calorie-restricted mice can live up to 50 per cent longer. The benefit in the monkeys was much more modest, probably in the range of five per cent to 10 per cent. The lessons for humans from this are not entirely obvious. Assuming that CR does work it's far from clear when you should start it, what the diet should consist of, and what happens if you fall off the wagon and have the occasional bar of chocolate.

It is quite hard to get to the bottom of what is actually going on. There are many factors involved in the response to diet apart from the actually calorific intake. These include the quality of the source of calories, the pattern of feeding, fat composition and specific nutrients in the diet. One hypothesis suggests that CR primes cells by activating stress pathways. Once activated the cells can cope better with subsequent cellular stresses. In this scenario cells move from active growth and proliferation to a state that favours repair and maintenance.

Another proposed mechanism is through a beneficial effect on mitochondrial and antioxidant activity. With fewer calories in the diet there are changes in the body's metabolic balance, including in the insulin and growth factor systems. CR has also been shown to alter gene expression and the pattern of epigenetic markers. There is, as yet, little consensus over the impact of these changes on longevity.

In short, fewer calories in the diet appears to be associated with more efficient metabolism and greater protection against cellular damage. Cardiovascular disease and type 2 diabetes are age-associated diseases. Diet and exercise have been shown to delay onset and progression of both these diseases.

Results from small human trials show that CR is associated with beneficial metabolic changes. Another small study suggested that regular exercise had much the same effect as CR, at least in changes in body composition and fat distribution.

A larger study called CALERIE (Comprehensive Assessment of the Long-term Effects of Reducing Intake of Energy) investigated the effect of restricting caloric intake for two years. Some 200 non-obese volunteers were randomised to have a normal diet or one with 25 per cent fewer calories. Given the timescale of both the study and a human lifespan it was not possible to look at whether longevity was increased. Instead the outcomes were changes in metabolic rate and core temperature, and various chemical, physiological and psychological measures. Not all the participants stuck to the goals. Overall, 82 per cent of those on calorie restriction managed to carry out the protocol. They only averaged 11.7 per cent fewer calories than controls, and that still meant they were eating around 2,200kcal per day. Baseline weight averaged about 80kg for the 66 men in the trial and 68kg for the 152 women, with average BMIs of about 25. Overall, those on the diet lost about 7kg. Remember the participants were from the United States. Although they were perhaps unusual in not being fat to start with, the amount of weight loss hardly put them into an emaciated category. The metabolic changes were minor and don't really provide much proof that calorie restriction will increase longevity in humans. They probably did demonstrate that this amount of calorie restriction does no harm, but I wouldn't expect it to in the people studied.

Intermittent fasting is one way of restricting the intake of calories and may be more acceptable over a long period than a daily diet.

Additional benefits have been claimed including the effect of intermittent nutritional stress on cellular function and repair. The 5:2 diet is one example of this regimen. An alternative is the "fasting mimicking diet" (FMD) that involves cutting calorie intake for four consecutive days a month. In diabetic mice this diet restored function in insulin-producing cells reversing their diabetes.

There has been criticism of some of the animal studies. The control groups are often allowed to eat freely. The findings may thus be simply demonstrating the shortened lifespans among the gluttonous creatures allowed to eat what they want compared with the CR animals who can't get at the research equivalent of chocolate and crisps. Although CR-related longevity is common it is not universal. In different strains of mouse longevity can be decreased, increased or unaffected suggesting that the effect depends on the genotype. Furthermore, there is still considerable uncertainty whether any beneficial effect is seen in humans.

An interesting, but rather off the wall, finding is that eating "old" food may affect lifespan. The hypothesis is that older organisms will have more cellular damage and if eaten may cause the animal eating them to age faster. To test this theory yeast were fed on cultures from either old or young yeast, fruit flies were given a feed based on old or young flies, and mice were fed meat from old or young deer. Most of the organisms on the "old" diets had a shortened lifespan. Perhaps old organisms are not nutritional as good, or perhaps old tissues contain harmful substances.

After all this evidence if you want to live a long life should you adopt a calorie-restricted diet? Are you the sort of person who enjoys self-denial and leads an ascetic lifestyle? If so, why not? It's unlikely to do you any harm and will save you lots of money. But would you spend it on enjoyment anyway? It won't stop you dying from being hit by a bus. However, it may well make you feel as though your life is longer as you wait between meals with a rumbling tummy. As for me, I'll carry on enjoying my food and alcohol, in moderation as always.

22 Extending life: Drugs and nootropics

"A man has only one way of being immortal on this earth: he has to forget he is mortal."
Jean Giraudoux (French novelist, 1882-1944)

Experiments in animals have shown that many interventions can increase lifespan. That provides hope that there may be drugs that can work in humans. There are many chemicals that have potentially beneficial effects on cells in culture or, in laboratory conditions, in bacteria, yeasts, rodents or even primates. To prove that anything is effective in humans is going to be hard. After all, there are ethical constraints about keeping people in laboratory conditions. We live a long time and waiting for results as people die off is going to be tedious as well as expensive. There just might be problems with the relatives if the life-prolonging intervention turns out to be harmful, or even from those given the inactive placebo if the drug works well. Anyway, humans don't make good research subjects. They forget to take the drugs, they have a craving for a hamburger and ruin the dietary protocols, they take an array of other interfering medicines and dietary additives, decide they don't want to participate anymore, go on holiday at a crucial time or fall under a bus thus playing havoc with mortality figures.

Geroprotectors are substances that will slow ageing, repair age-associated damage and extend healthy lifespan. As of 2017 more than 250 of them had been investigated in a wide variety of different models of ageing. Developing a new drug typically costs billions of dollars and can take 10 to 20 years to take it from early-phase testing to the market. It is therefore much cheaper and quicker to look at drugs that have already been approved for other indications to see if they have any role as anti-ageing agents. More than 100 potential geroprotectors are already approved for human use and there is considerable interest in repurposing several of them. To date there is no anti-ageing medication approved for use in humans. Nonetheless there is a range of drugs that have promise and are being investigated. The National Institute on Aging (NIA) website (accessed January 2017) has a list of compounds currently being tested in mice. These include rapamycin, minocycline, aspirin, insulin, acarbose, metformin, fish oil, curcumin, green tea extract, simvastatin and resveratrol.

Calorie restriction (CR) is one intervention associated in some models with a longer lifespan (see chapter 21). CR affects many systems including energy metabolism, insulin sensitivity, and inflammatory and neuroendocrine responses. It also down regulates TOR (as in Target of Rapamycin) function leading to more controlled cell destruction and a decrease in protein synthesis. Wouldn't it be good to have a drug that mimics the effect of CR without having to diet? That is one of the main drivers behind some of the medication being investigated. Compounds that mimic CR do so by altering cellular function. Many of the drugs with these effects are naturally occurring compounds. Here are some examples.

Rapamycin was first derived from a bacterium found in the soil of Easter Island. The Polynesian name for the island is Rapa Nui, thus the drug name. It has been used clinically as an immunosuppressant following organ transplantation. It appears to mimic the effect of CR inhibiting a ribosomal enzyme called S6K as well as TOR itself. These regulate protein translation, increase growth and inhibit cellular autophagy. In a range of organisms including yeasts, nematodes, flies and human cells it has been shown to increase lifespan. It was also the first drug shown to increase lifespan in mammals, lengthening by 25 per cent the median time mice lived. The effect is related to the dose of drug given and it is possible that an even larger dose will result in greater longevity. Not only do mice live longer but age-associated changes in many organs are delayed in onset. Most mice die due to cancers. It is possible that the longevity is because of an anti-cancer rather than an anti-ageing effect. Rapamycin is already approved as a drug to prevent organ rejection after kidney transplantation and to prevent cardiac stents blocking. Giving it to human subjects should therefore not be a major barrier. Pet owners are being asked to enrol their dogs in a study to see if it combats the effects of age and extends lifespan. Human studies will not be far behind.

Polyphenols are produced by plants in response to stress or fungal infections. They are free radical scavengers and also have the ability to inhibit inflammation. These compounds produce, at least partially, the same effects as CR. Among these compounds, **resveratrol** has emerged as the one that best mimics the effects of CR. However, no clear benefit on lifespan was found in studies where resveratrol was given to fruit flies and nematode worms. It is commonly found in a variety of foods including berries, peanuts and grapes. In red wine it has been linked to a decrease in heart disease. It may also have some effects that will prevent or treat cancer. Given to healthy obese men it reduces sleeping and resting metabolic rate and has a range of measurable effects on metabolic activity similar to that seen with CR. At the genetic level it may interfere with pathways responsible for inflammatory signalling and for DNA damage. Unsurprisingly, some people who live in the winemaking Chianti region of Italy ingest large amounts of resveratrol in their diets. During a nine year follow-up of elderly residents there was no discernable effect of resveratrol intake on health or mortality rates. This raises as many questions as it answers. Was there enough resveratrol in the diet to make a difference? Were the methods of estimating dietary intake robust enough and were the outcome measures valid? Perhaps the adverse effects of red wine balanced out the good in resveratrol. It may be worth noting that the researchers stated that there is "limited and conflicting human clinical data demonstrating any metabolic benefits of resveratrol, and there is no data concerning its safety in high doses or for long-term supplementation in older people, who often have multiple comorbidities for which they are taking multiple medications". In the meantime it's probably safe to keep drinking Chianti in moderation (notwithstanding advice from the Chief Medical Officer).

Drinking large quantities of **green tea** has been given as the reason why some Japanese live to an extremely old age. One of the ingredients of green tea is **catechin**, another natural phenol. It is also a free radical scavenger and may inhibit the inflammatory response. It has potential as an anti-cancer agent by promoting cell death in tumours.

Drugs that block **glycolysis** (a metabolic process that breaks down carbohydrates and sugars to release energy for the body) may mimic the physiological effects of CR. One of these, **2-deoxyglucose**, has been shown to increase the lifespan of simple organisms. However, it may cause heart damage and increase the risk of cancer so is unacceptable for human use.

Metformin is a compound derived from French lilac. It is currently widely prescribed as a drug to manage diabetes. It has been shown to increase the lifespan of some bacteria, worms, mice and other species. In low doses it appears to produce similar effects to that of CR. It can restore mitochondrial function in old cells in a culture medium. In human studies it has been shown to reduce the chances of getting type 2 diabetes. It has also been shown to lessen the risk of developing heart disease. There is also some evidence to suggest it may decrease the incidence of cancer and also cancer-related mortality. This wonder drug may also preserve cognitive function. Metformin has been safely used for many years so there is no major barrier to its investigation in human subjects. A study called TAME (Targeting Aging with Metformin) has been planned. This proposed trial marks a huge shift from treatments aimed at specific diseases to one aimed at the human condition itself.

How effective metformin will be remains to be seen but it does illustrate the focus of future research.

Arcobose is an α-glucosidase inhibitor used to treat diabetes. Mice treated with it live 22 per cent longer. As well as its effect of reducing blood glucose concentrations it can decrease blood pressure, and lower the chances of heart problems.

Deprenyl is an antidepressant that prolongs the lifespan in male rats.

Spermidine is yet another polyamine that can be found in the cells of many organisms. Although they are naturally produced within cells, dietary ingestion can alter gene expression and chromatin organisation that is normally associated with increased health and longevity.

Aspirin (acetylsalicylic acid) is a non-steroidal anti-inflammatory drug that was previously used widely as a painkiller. Through its action on platelets it also makes blood less sticky and inclined to form clots. More recent investigations have focused on its potential to destroy cancer cells.

We have seen that epigenetic factors play key roles in controlling changes in gene expression. Interestingly, drugs that modulate the activity of certain enzymes can induce epigenetic changes. DNA methylation inhibitors may be active against various cancers and may also be used to treat psychiatric diseases such as schizophrenia and bipolar disorder. **Azacytidine** and **decitabine** have these effects although toxicity and poor chemical stability restrict their potential uses. Histone deacetylase inhibitors have been used to treat cancer and heart problems. Some compounds have already been used in human clinical trials as anti-cancer drugs or for treating age-related diseases. **Vorinostat** and **romidespin** are histone deacetylators that have been approved for the treatment of lymphomas. Drugs of this class may have a role in the treatment of neurodegenerative disorders such as Parkinson's disease, Huntington's and Rett syndrome. **Sirtuins** produce a compound that encourages cells to repair their DNA. Claims have been made that, in mice, it can reverse the human equivalent of three decades of ageing. Human studies were said to be starting in 2017

If cells continue to survive in the body they become senescent (see page 63). Inducing senescence can be desirable, for example, in treating cancer. Here radiotherapy and chemotherapy target cancer in an attempt to selectively stop these fast-growing cells from developing. Preventing the natural accumulation of senescent cells may contribute to healthy ageing and extend lifespan. The class of drugs that target and kill senescent cells are called **senolytics**. One such drug is **dasatinib** and another is **quercetin**. Both have been used in humans for other indications. Even if senescent cells can be removed it is far from clear how and when to initiate treatment. Although the potential is obvious it is not known whether there will be benefits in terms of disease progression or increased life, and there are considerable theoretical risks of inducing harmful pathology.

There are several major neurological conditions where there are abnormal amounts of protein. These include Alzheimer's, Parkinson's and Huntington's Disease. One line of enquiry is based upon the possibility that the underlying metabolic mechanism is the production of oxygen free radicals. There is increasing interest in foods that may be able to inhibit, delay or even reverse the steps leading to neurodegeneration. **Curcumin** is an ingredient in turmeric and **ferulic acid** is found in fruits and vegetables. Both may reduce oxidative damage and decrease the generation of amyloid plaques that are associated with Alzheimer's.

Several interventions have been shown to activate telomerase and thus prevent telomeres shortening. These include **cycloastragenol** (also called TA-65) that can be bought as a herbal food supplement. It is an extract of *astragalus propinquus*, which is apparently one of the 50 fundamental herbs used in Chinese traditional medicine. One of the ways resveratrol may work is through activating telomerase. **N-acetylcysteine** is an antioxidant which blocks transport of telomerase out of the nucleus. It has also been hypothesised that the mode of action of some antidepressants and other psychoactive drugs may, at least in part, be mediated by their effect on telomerase activity. A major concern about enhancing telomerase activity is the risk of cells becoming cancerous. But don't forget normal human activity such as exercise is associated with an increase in telomere length. A therapeutic approach that may mitigate the risk of developing cancers is being developed in which telomerase is induced temporarily and selectively in old cells. Telomerase has been targeted in several anti-cancer treatments. In this case the idea is to depress the enzyme to stop cells dividing.

The Hutchinson-Gilford Progeria (HGP) syndrome is caused by the generation of progerin instead of lamin A (see page 72). Several compounds (**JH1, JH4** and **JH13**) have been shown to block binding of progerin to lamin A. In cell culture these drugs reverse some of the signs of ageing. Furthermore, when given to a mouse model of HGP they decrease some of the signs of premature ageing and extend lifespan. Trials in human subjects are taking place of drugs including **pravastatin, zoledronic acid** and **lonafarnib** that inhibit the metabolism of progerin. Early results in small numbers of patients have demonstrated some minor improvement in bone mineral density. There are some hints that the rate of ageing can be reversed. One intervention involves depleting methyltransferase that has the effect of improving DNA repair, delaying the signs of senescence and extending lifespan in a mouse model of progeria.

Other metabolic pathways will undoubtedly be investigated. There are already interventions that will increase healthy cell production. The use, or rather abuse, of **erythropoietin** (EPO) is well known because it has been taken by cyclists and endurance athletes to improve oxygen carrying capacity. It has a legitimate medical role in treating anaemia, particularly that associated with kidney failure. Other growth factors include granulocyte colony stimulating factor (**G-CSF**) and granulocyte-macrophage colony stimulating factor (**GM-CSF**). Energy control is central to healthy cells. AMP-activated protein kinase (**AMPK**) regulates energy balance and is activated in response to conditions that deplete cellular energy levels. Mutations in its structure have been shown to affect some disease states.

Go to the webpage of **nootropics.com** to find "the world's most powerful, effective and safe cognitive enhancers". These are their words not mine. **Nootropics** are substances claimed to enhance memory and learning while lacking any sedative, stimulant or toxic effects. They're not medicines to treat illness but rather substances to make you perform better while still healthy. The website is really an ecommerce site selling pills for tens of dollars a tub.

What happens if you become super brainy is a common subject matter in science fiction. In the cinema Eddie Morra (played by Bradley Cooper) in the 2011 film *Limitless* takes the fictitious nootropic drug NZT-48 that transforms him into an intellectual superman. The drug CPH4 gives Lucy (Scarlett Johansson) enormously enhanced mental and physical powers in the 2014 film *Lucy*.

The claim is that not only will nootropics boost productivity, energy and motivation, but they will provide confidence in social circumstances with the implied promise of being able to finally land a girlfriend or boyfriend. The list of nootropics and claimed benefits is long, covering everything from anti-anxiety to vitality. The following are examples of three of the products from the 10 shown on the website at the time of writing. It appears to be quite common to combine several nootropic compounds and this is called stacking. The phrases in quotation marks are from their literature.

L-Theanine: "Has anti-anxiety and stress-relieving properties giving a clean, lucid energy boost". This is an ingredient of tea. The capsules on sale at nootropics.com have about 10 times the amount that is in a cup of tea. They suggest that one to three capsules are taken a day.

Piracetam: "Improves focus, memory, and verbal fluidity". It has been used clinically for the treatment of dementia and cognitive impairment. A comprehensive review said there was a need for more research although the limited published data does not support the use of this drug.

Adrafinil: "When adrafinil is metabolised to modafinil in the liver it produces a state of heightened cognitive ability characterised by concentration, wakefulness, and focus". Adrafinil is a prodrug that needs to be metabolised in the body to form the active ingredient modafinil (a prodrug is an inactive substance that is converted to a drug within the body by the action of enzymes or other chemicals). There is little information about adrafinil and modafinil is the main, but not the only, metabolite. Modafinil is used to treat sleep disorders. There is a risk of developing hypersensitivity and serious adverse psychiatric reactions, and it may be addictive.

There is no doubt that some substances stimulate the central nervous system. A coffee shot is essential to many people for just that reason. Other well-recognized centrally acting stimulants include amphetamine and nicotine. Before regularly taking a drug that affected my brain I would need to be convinced that there was good information to support its effectiveness, and also its long-term safety.

Some drugs being investigated for their anti-ageing properties:
2-deoxyglucose; acarbose; aspirin; azacytidine; catechin; curcumin; cycloastragenol; dasatinib; decitabine; deprenyl; ferulic acid; fish oil; green tea extract; insulin; lonafarnib; metformin; minocycline; N-acetylcysteine; nicotinamide mononucleotide; pravastatin; quercetin; rapamycin; resveratrol; romidespin; simvastatin; sirtuins; spermidine; vorinostat; zoledronic acid

---

23 Extending life: Gene therapy

"I don't want to achieve immortality through my work. I want to achieve it by not dying."
Woody Allen (From *The Illustrated Woody Allen Reader*, 1993)

The gene genie is truly out of the bottle and gene therapy is now an established technique. If you have £50,000 or more to spend, and want to recreate a much-loved pet dog, then it is already possible to pay to have it cloned. In a technique called "somatic cell nuclear transfer" the nucleus from a skin cell of the animal to be cloned is inserted into an egg from which the nucleus has been removed. The egg is then placed in a surrogate mother where it grows to be born as a replica of the original. However, the clone is emphatically not the original animal. It is a copy, and not even a perfect copy. Obviously, it lacks the experiences, memories and habits of the original. Although the genome may be identical it will have different epigenetics and therefore subtly different gene expression. Nonetheless, cloning techniques are already being used to make pampered pets, and it can also be a way of saving and preserving endangered species.

There is no practical barrier to stop humans being cloned. Your twin or replica could live on in your genes The possibility has been the plot for several books and films. It raises fascinating questions about nature and nurture. How much of an individual's characteristics is due to his or her genome and how much is determined by their environment and upbringing? Clone Adolph Hitler and what are the chances of raising a genocidal mass murderer or a house painter?

Cloning is not the same as living an exceptionally long life. The genes may live on but they survive in a completely different individual.

At least 4,000 genes have been associated with various human diseases and many more are likely to be discovered. There is also a clear genetic component associated with longevity. The GenAge database lists genes thought likely to affect the length of life. All these genes have been shown to increase lifespan in one or more of a wide variety of organisms. Depending on the gene this might be when they are removed, inserted or overexpressed. In early 2017 the website listed 2054 genes in various models (yeast, worms, flies, mice etc.), 305 of which were thought to be important in humans.

In some inherited conditions a family history suggests that parents are passing on genes causing the disease. Prenatal screening can then be used to identify foetuses carrying these genes and the pregnancy terminated. It is now becoming feasible to screen for defects before conception. An egg without the genetic defect can be selected. It is then inseminated and implanted in the mother's womb using IVF procedures, thus guaranteeing a foetus free of the specific genetic defect.

In other cases genetic alterations occur by spontaneous mutation. These are not inherited but are alterations in base pairs of the genome, usually an addition, substitution or deletion. There may also be failures in DNA dissociation and amalgamation. In Down's syndrome, for example, there is a failure of DNA separation resulting in three, rather than two, copies of chromosome 21. Although babies with Down's are born to mothers of all ages, the chance of this happening goes up with the mother's age. In a 20-year-old it is about one in 1,500 births, while for mothers aged 45 it is one in 50 births. This is one example of how external factors affect the likelihood of conceiving a child with a genetic abnormality. A female has two X-chromosomes and males an X and a Y. Some genetic defects are carried on the Y-chromosome only. If this is the case only a boy can inherit the disease.

Although single point mutations cause syndromes with many of the features of ageing, the genetic contribution to human ageing appears to be diffuse and ill understood, and may be partly determined by accidental and random effects. Any genetic modification to increase longevity is likely to be complex and fraught with possible unexpected ramifications. Despite that, over the last 20 years a large number of genetic manipulations have been shown to extend life in a range of invertebrates.

The Methuselah Mouse Prize is awarded for creating the world's oldest mouse. Individual mice rarely live past two years of age and these unfortunate creatures bear the burden of much of the animal work into longevity. In the laboratory they've been selectively bred, had genes knocked out and put in, been starved and drugged, poked and prodded, studied and sacrificed, all in the name of science. Has it been worthwhile?

As of early 2016 the lifespan of a mouse had been extended to about 4.5 years. It is results such as this in a mammal that gives hope to those searching for a breakthrough in efforts to improve human longevity.

Scientists have now reached the stage where it is possible both to read and to write genes. Recombinant DNA is the joining together of two DNA segments that are not naturally found together. This involves the use of enzymes that can cut, link and modify DNA. DNA sequences can be designed and created and then multiplied. They can then be inserted into the DNA of the organism to be altered. A virus is used to carry DNA into affected cells thus altering the recipient's genotype. Viruses consist almost entirely of DNA or RNA. They enter, infect a cell and then use the cell's mechanisms to replicate the virus genetic material.

When it comes to gene editing it's worth learning about **CRISPR** or Clustered Regularly Interspaced Short Palindromic Repeats. Got that? It has been adapted from a bacterial system for defence against foreign invading genetic elements. It is a genome editing system, the elements of which are non-coding RNAs and Cas9 (CRISP associated) proteins.

Potentially this technology can go to any strand of DNA from any life form, snip it in a precise location and then add or remove DNA.

The basic CRISPR system has been refined and modified to improve its function. There are still, however, technical issues that need to be overcome to make it reliable and accurate. Nonetheless, it is an extraordinary tool and has the potential to transform genetic engineering. It is low cost and easy to use and has only been around since 2012. So far it has been used extensively to modify cultured cells *in vitro* in order to test drugs. Gene editing to correct mutations has also been tried with some success in viable human embryos. It is already starting to be used to alter cells in humans. *The Times* (February 19, 2017) reported that a two-year-old girl with leukaemia may be the first person in the world to have been cured through gene editing. She was given modified donor T-cells targeted against the cancer.

As well as gene editing it is now becoming possible to edit the epigenome – the bits of DNA that control and modify gene expression. The system uses the same targeting as in gene editing. Epigenetic regulation can occur through methylation and demethylation of DNA bases. Chromatin is the combination of DNA and protein that makes up the content of the cell nucleus. Histones are the main protein in chromatin and modifications in histones affect gene expression. Epigenetic changes are important mechanisms for regulating gene expression and they may be an important mechanism for controlling ageing. Changes in chromatin have been identified in old cells, changes that may be a response to environmental stress. It is therefore believed that epigenetic alterations occur as a response to the ageing process but may also be a cause of ageing. Reversing ageing may have risks and uncertainties, in particular the possibility of developing malignancies in cells that have acquired genetic mutations during ageing but have gained greater potential to proliferate through rejuvenation.

The first molecule of recombinant DNA was created as recently as 1971. In 1974 a frog gene was inserted into a bacterial cell. Man had succeeded in doing something that was impossible in nature. Initial trials of gene therapy were disappointing because of unwanted immune responses and the increased incidence of cancer. Many practical problems still remain but are slowly being addressed. Scientists can now manipulate any chosen gene and incorporate it into the genome. The ability to do this is astonishing.

To change a single base is the equivalent of taking 850 volumes of the complete works of Shakespeare and purposefully editing one specific letter. It's an extraordinary task to even contemplate and it's even more amazing because it can be done.

In 1978 insulin was one of the first substances to be genetically engineered. As well as manufacturing insulin, recombinant DNA is already being used extensively to produce safer vaccines, growth factors, food additives, blood clotting factors and diagnostic tests. Gene therapy is now a reality and mainstream. Treatments are already being successfully administered and many more are in the pipeline. Strategies for clinical gene therapy are either to change the gene within the patient's body or to introduce cells altered outside of the body and reinfused. Treatments have been trialled in several conditions including primary immune deficiencies, haemoglobinopathies, haemophilia B, cancer and some neurological and eye diseases.

Children with severe immunodeficiency are often called "bubble babies" because they live in a germ-free environment as their immune systems cannot fight infection. In one form of this disease patients have a mutation in a gene encoding adenosine deaminase (ADA), an enzyme responsible for clearing toxins from the body. Gene therapy inserts a corrected copy of this gene. In some children this treatment has returned them to normal, an extraordinary result for a devastating disease. ß thalassaemia is caused by reduced or absent ß chains in haemoglobin, the molecule that carries oxygen in the red blood cell. It is common and causes many serious complications. Conventional treatment consists of frequent blood transfusions and that is unsatisfactory, expensive and has unwelcome side effects. An engineered ß globulin gene has been inserted in several patients markedly reducing the need for transfusion.

Haemophilia B (sometimes called Christmas disease) is caused by a mutation causing deficiency in clotting factor IX. Gene therapy increases blood levels of factor IX. A single treatment keeps levels raised for at least three years and reduces bleeding episodes by more than 90 per cent.

Sickle cell disease affects millions of people worldwide. They have an abnormal version of haemoglobin. In 2015 bone marrow stem cells were obtained from a patient, a 13-year-old French boy. A normal gene was inserted into them and then they were retransfused. Fifteen months after treatment he was producing normal haemoglobin and was off all medication.

In 2017, for the first time, a personalised gene therapy for cancer was approved for use in humans. This treatment for acute lymphoblastic leukaemia involves taking T-cells from the patient (T-cells are a key component in the body's normal immune response). These are then genetically modified so that they will recognise proteins on the surface of the tumour cells. They are then reinfused into the patient where they mount an immune response against the cancer. Preliminary data suggests that long-term remission can be expected in about 80 per cent of those treated.

Millions of people are on statins to lower blood cholesterol. Individuals who have a mutation preventing their liver making a protein called PCSK9 have naturally low cholesterol levels and a much lower incidence of cardiovascular disease. Using CRISPR the relevant PCSK9 gene has been disabled in mice holding out the promise that gene therapy may be effective in humans lowering cholesterol naturally without the need for lifelong medication [report from a conference – *New Scientist,* February 2017].

Up to now gene therapy has focused on easy targets where the disease is caused by the absence of a functional protein and a cure can be achieved by the addition of a specific gene. Attention is now being increasingly turned towards gene editing, correcting disease-causing genes or targeting a therapeutic gene into a defined genetic locus.

Optogenics is the process of inserting a gene into cells to make a light sensitive protein. Shining a laser on the treated cells makes them alter their behaviour in various ways that depend on the protein produced. This could potentially be used as a targeted treatment against cancer and it has already been trialled as a treatment for the eye disease *retinitis pigmentosa*. In this study, blind people are injected with a virus containing DNA from a light-sensitive alga. The hope is that this DNA will get into the ganglion nerve cells enabling them to send visual information to the brain. In another experiment, mice were infected with a virus that made the neurons in their brains sensitive to blue light. When a blue laser was shone on the amygdala area in the brain it triggered hunting behaviour.

Experiments like this improve our understanding of the way the brain works. It also shows how behaviour can be controlled. What

happens to free will (if we ever truly had it) when how we act and what we think are at the whim of reactions taking place deep in our brains over which we have no control?

Mice given gene therapy to increase telomerase activity live much longer and have better health and fitness. Telomerase expression has also been induced in human cells in culture. This has the effect of enormously increasing cell lifespan while maintaining a youthful appearance. Genetically-engineered plants and animals have been created. Plants can be made resistant to certain insects or herbicides, sheep can produce clotting factors in their milk, and pigs can make human haemoglobin. A jellyfish gene has been inserted into a mouse so that it glows in the dark under blue lamps (why you may ask?). Also, in a mouse, inserting a gene controlling communication between neurons results in increased memory and better cognitive function.

Precise, targeted modifications have been introduced into chicken genomes to increase resilience to disease and replace chicken immunoglobulin genes with human sequences. Jellyfish DNA has even been added to the male chromosome of chickens. The eggs holding male birds can then be made to produce an easily-spotted green fluorescent protein. Unwanted males are then weeded out and destroyed before hatching.

A pig is similar in many ways to a human, although probably not in the ways that first spring to mind. It's their relative anatomy, physiology, organ size and cell cycle characteristics that make them interesting. Intriguingly, there is apparently a larger evolutionary distance between humans and pigs than between humans and mice. The disadvantage of mice is they're the wrong size. Pigs can be engineered so that organs taken from them do not evoke an intense immune response when transplanted into other animals. Hearts and kidneys from these pigs have been placed into non-human primates who have subsequently survived for several months. This is called a **xenotransplantation**. Liver transplants have also been attempted with much less success.

A step beyond xenotransplantation is to grow human organs in another animal. Genetic chimeras have been created combining the genes of different mouse species, and by incorporating bits of rat genome into mice to produce rat pancreases. This raises the possibility of generating working human organs in other animals. We obviously breed pigs just to eat them. Is it any different to breed them solely to harvest organs for transplantation?

If you worry about animal rights how about cloning your own human as an organ reservoir? A world has already been imagined where clones are created to be organ donors. The 2005 film, *The Island*, and the 2010 film, *Never Let Me Go*, are just two examples in fiction where boys and girls have been cloned with this specific purpose. It would probably seem much more acceptable and ethical to design out the troublesome thinking and feeling parts of the brain reducing these humans to the level and status of animals.

It may be macabre to think about but if you or I had a cloned copy of ourselves, minus the cerebral cortex, wouldn't we have an excellent supply of spare parts?

This would provide organs that can be plugged in without any problems of rejection to replace failing kidneys or hearts or livers or intestine or any other bodily part that can be removed and replaced with a bit of simple vascular surgery joining up the arteries and veins that provide the blood supply and the venous drainage. Without the cerebral cortex the clone would not be conscious, have no memories, no volition, no emotions, no sense of self. But with an intact brainstem there would be the necessary central nervous control to ensure that the lungs worked, the heart beat and the body, more or less, was able to survive. It's not quite that straightforward because nutrition would have to be administered, regular turning to prevent pressure sores would be essential and intensive nursing would be needed to keep the body clean and functioning. If society is willing to tolerate these living, breathing organ repositories it is only one small step further to creating gruesome monsters who can function enough to live independently with the instincts and intelligence of a cow or sheep, and are reared only to be slaughtered when the need for organs arises. There are pretty obvious ethical and moral objections to this that make it completely unacceptable. Growing clones without brains may get round this problem but is macabre, twisted and probably not very effective.

Growing isolated organs would be a much more acceptable solution and may also be feasible in time.

A lot of genetic research has involved adding (knocking in) or removing (knocking out) genes. By breeding altered animals and plants insights are gained into the role of the affected gene. It has been estimated that only about 0.025 per cent of the biological molecules active in the human body are, at present, targets for therapeutic intervention. With knowledge of the genetic basis of these molecules, the potential for developing effective, targeted and even personal therapeutic interventions is obvious. There is no practical reason why the cell germ line cannot be edited. However, this is much more controversial as those changes will be able to be inherited and passed onto future generations.

Altering the genotype of an animal or plant generates concern from those worried that out-of-control harmful genes may spread. There are also moral and ethical concerns about fundamentally altering the makeup of a living organism. These concerns should not be ignored and need to be addressed. Selective breeding has been used to alter genes in animals and plants since the dawn of civilisation. It is a different matter when selectively breeding our own species because of unacceptable overtones of ethnic cleansing or racial purity.

However, recombinant DNA technology has the potential to prevent or treat many illnesses and it will undoubtedly be an important medical intervention in the future.

Life is based on the DNA software. Change the DNA software and you change the species. Take the genome out of one species and put it into another animal's cell and you convert one species into another. The code of life has also now been digitised. Scientists send each other digital code instead of genes or proteins. It is faster and cheaper to synthesise a gene than it is to clone it. If an email is sent with a digital sequence for, let's say a flu virus, it can be turned into the genome in less than 12 hours. We are on the cusp of building synthetic life. Mars could be colonised in the few minutes it takes to send digital information rather than the dangerous months a space ship would take.

Completing the sequencing of a human genome was a milestone in human biology. Having done this the next challenge is to identify the structure of all functional elements in addition to the genes; in other words to find out what every bit of the genome does. Nearly 99 per cent of the 3.3 billion nucleotides do not code for proteins. However, large chunks of non-coding genome are important through regulation of gene expression. They can be looked at using three different types of analysis. Firstly, what is the effect on the phenotype of altering the genotype? Secondly, which elements of the genome have been selected and conserved during evolution and are thus likely to be important? Thirdly, what do the genes do at the biochemical and cellular level? All three approaches are useful. For example the effects at the level of the cell may depend on the cell type whereas the evolutionary pressures depend on environment and evolutionary niche.

In September 2003, the National Human Genome Research Institute (NHGRI) launched a public research consortium named ENCODE, the Encyclopedia Of DNA Elements, with the goal of identifying all functional elements in the human genome sequence. A pilot phase tested and compared existing methods to analyse rigorously a defined portion of the human genome. The findings demonstrated that it was possible to identify and characterise functional elements. New technology was developed and tested so that data could be more easily gathered. A data coordination centre can be accessed online (**genome.ucsc.edu**) to see the encyclopaedia of DNA elements. Preliminary data suggest that about 11 per cent of the non-protein coding DNA is associated with regulatory proteins. About 20 per cent is associated with enhancers or promoters of histone, and a third is associated with control of transcription. It should be stated that not everyone agrees with this analysis and others believe that only eight to 14 per cent of DNA is likely to have a function.

Although considerable progress has been made, there is still much to do before we truly understand the whole human genome.

A whole new science and industry of biotechnology has been created in the last forty years. Synthetic biology is where science and engineering approaches come together to precisely control biological networks. This approach means that artificial genes can be designed and made for implantation into a host cell. To date genetic engineering has been limited to small specific changes in the genome. In the future it may be possible to make novel proteins and genetic circuits that have no direct analogue in normal human biology. DNA has already been constructed using novel base pairs not found in nature. The vision is one of artificial chromosomes as functional additions to the normal human genome. A complete bacterial genome has already been constructed from scratch so creating artificial chromosomes is essentially a problem of scale. These artificial chromosomes could be made either by building up from basic components, or by stripping down a natural human chromosome. Although this may sound like science fiction it isn't.

Cell implants with embedded synthetic gene networks have already been used in mice for the treatment of gout, thyroid dysfunction and obesity.

Microbes can sense a wide variety of stimuli, process information and produce a range of chemical and physical responses. Software can already design DNA sequences to build biological circuits for functions such as decision-making, control and sensing. New biological tools are being created to engineer cells to become cellular foundries and nano factories. Genetic engineers have reprogrammed cells to produce desirable compounds, sense and report on the presence of toxic metals and seek and destroy pathogenic bacteria.

Biohacking, the low cost, do-it-yourself offspring of biotechnology has also been spawned. The fall in the cost of sequencing DNA allows amateurs to duplicate at home techniques that used to be the preserve of the government and large corporations. Standardised chunks of DNA can be bought off the shelf and laboratory equipment is no longer beyond the budget of hobbyists. The potential to design and build living organisms is limited only by the scope of the biohacker's imagination. Some commentators have compared the growth of home biotechnology to the early days of computing back in 1975 when Apple built their first computer. And we all know what has happened to Apple since then.

Mankind stands on the threshold of a revolution in the prevention and treatment of disease, the benefits of which are only just becoming a reality. We will have the potential to change the biological world around us with genetically modified plants and animals. We may soon have the means to change and enhance the human genome in a way that would be impossible in nature.

24 Extending life: Regeneration

"Children are the only form of immortality that we can be sure of."
Sir Peter Ustinov (actor, wit and author, 1922-2006)

An understanding of the factors that modulate regeneration suggests strategies for delaying ageing and for renewing tissues. With regenerative medicine the body is enabled to restore functionality using its own restorative powers. Medical science is only just beginning to understand and apply the insights that have been gained. Exciting discoveries are being made and translated into novel treatments.

The discovery and use of **stem cells** is already transforming medical therapy. There is little doubt that they will become even more important in interventions to treat disease and prolong life. As we have seen (page 66) it is extraordinary that any somatic cell in our body can be turned back into a pluripotent stem cell. This is a cell that is undifferentiated and can then be made to turn into any other cell. Stem cells have been classified into unipotent, multipotent, oligopotent, pluripotent or totipotent depending on the number of cells into which they can be differentiated. Totipotent cells can form embryonic as well as extra-embryonic tissue such as the placenta. Pluripotent cells can differentiate into other cells of the adult body. As cells differentiate they lose their ability to turn into other cells. In theory induced pluripotent stem cells (iPSC) can be generated by any somatic cell. There are several methods for introducing reprogramming factors into cells in culture. One way is through viral vectors, another by directly adding reprogramming proteins.

Reprogramming of somatic cells into iPSC is the ultimate **rejuvenation**. It is much more than the equivalent of a cellular shot of Botox or plastic surgery. Old cells look old with shortened telomeres, DNA damage and mitochondrial dysfunction. When senescent cells or those from centenarians are coaxed back to iPSCs they regain youthful characteristics. Telomere size is reset, gene expression renewed and the signs of oxidative stress and mitochondrial metabolism are indistinguishable from human embryonic stem cells. There is, as yet, conflicting evidence about whether the epigenetic memory is also reset.

Multipotent stem cells have been derived from a variety of tissues including cells from bone marrow, muscle and fat. Fibroblasts are cells that produce collagen forming the structural framework of animal tissue. They are about as far from brain cells as it is possible to get. However, only three genes need to be altered to change a fibroblast into a fully functioning neuron. Scientists have largely worked out the key signalling pathways and transcription mechanisms that make neurons develop. Just a few years ago the idea of regenerating brain cells seemed an impossible dream. Today many believe that we are on the threshold of achieving just that.

The ethical concerns of deriving stem cells from embryos do not exist if adult cells are the source. Enormous progress has been made in being able to harvest these cells and to make them pluripotent. A large number of trials are currently investigating uses of stem cells in anything from generating bone and cartilage to renewing heart muscle. One example in patients with stroke is the PISCES (Pilot Investigation of Stem Cells in Stroke) study into the safety of injecting stem cells into the brain. So far, results are encouraging with few side effects and some improvement in function.

Another study used iPSCs to make dopamine producing neurons. They were then transplanted into monkeys with Parkinson's disease. The transplanted cells not only survived but formed connections and improved mobility. Clinical trials are planned to start in 2018.

There is optimism that two or more cell types can be combined to make complex organs like the liver, heart or even brain.

There are challenges in getting hold of enough viable cells and theoretical risks of lack of compatibility and developing cancer. Early results even suggest that, in the right environment, stem cells can self-assemble to form an embryo without needing an egg.

With increasing age the regenerative properties of adult stem cells deteriorate. In muscle, for example, this means that muscle regeneration becomes inefficient and leads to a replacement of functional muscle by fatty and fibrous tissue. This deterioration may be due to the accumulation of damage within the cell or age-related changes in the environment surrounding the cell. By altering the cellular environment it has been shown that old cells can be rejuvenated and new cells made to act like old ones. Stem cells are also proving useful for giving insights into a wide range of conditions. For example, stem cells from children with autism form fewer neuronal connections than controls while those with the hyper-sociable Williams syndrome have an abnormally high number.

All medical innovations seem to follow a trajectory with enormous optimism followed by doom and gloom, before the rightful place and effectiveness of the treatment is established. Within a few years I expect to see reports of amazing cures using these cells. This may well be followed by reports of side effects, complications and lack of benefit. It will all become much clearer over time. One possibility is generating a safe and plentiful supply of human blood suitable for transfusion. Early blood cells have already been rendered immortal. The challenge now is to prove they are safe and able to be manufactured at a workable cost.

Such is the promise of stem cell therapy that I would be surprised if it didn't establish itself as one of the major contributions to medical advancement this century.

In 2013/14 there were nearly 3,000 adult kidney transplants carried out in the United Kingdom while there were 5,590 patients on the waiting list. There are always too few kidneys available for those who need one. Although the success rate is very high, rejection is a major long-term concern. Researchers have found a possible way of overcoming both of these problems. Precursor kidney tissue was dissected out of 16-day-old embryo rabbits. When transplanted into adult rabbits over half of them started to develop into differentiated kidney tissue. Even more exciting some of the embryonic tissue was vitrified. This is a fast freezing technique that requires an anti-freeze cryoprotectant and storage in liquid nitrogen at minus 196⁰C. After three months the tissue was thawed and transplanted into adult rabbits where a quarter were successfully grown. Perhaps, best of all, because the transplanted tissue is so immature it doesn't appear to be recognized as foreign by the host. If this system can be developed it heralds the possibility of being able to satisfy needs for a nonimmunogenic organ transplant. If it works for kidneys it may work with other organs.

Bioelectricity, the electrical signal that passes between cells, has recently been recognised as having a role in telling cells how to grow and where to go. In experiments using bisected flatworms the flow of ions between cells determines whether a head or tail is grown. Wound healing is, at least in part, under the control of electrical fields and some diseases are known to be associated with malfunction in the flow of electrically charged ions through cell walls. Changing the flow of sodium ions and increasing bioelectrical forces has already been shown to make froglets regenerate limbs at well past the age when this would normally be possible. Humans lose almost all their regenerative powers while still in the womb. Bioelectrical information may, just possibly, be one of the ways that regeneration can be stimulated.

Short telomeres are associated with degenerative diseases, premature ageing and early death. Mice with artificially increased expression of telomerase have longer telomeres, lower molecular damage and live longer and healthier lives. These findings provide the impetus to trial therapies targeted at increasing telomerase. Several compounds can boost telomerase activity and have been looked at in animal models. When adult differentiated cells are reprogrammed to a more pluripotent state the telomeres are also fully rejuvenated. A major concern over their use in humans is the fear that cancer could be a side effect.

In times gone by there was much myth involving the rejuvenation of an aged person with the tissues of a youth, usually a virgin. The idea that there is some youthful elixir in the blood is not totally far fetched. **Heterochronic parabiosis** is the surgical joining together of the circulations of two animals of different ages. When two mice are joined in this way survival is improved in the individual with a problem such as radiation damage or muscular dystrophy. In more recent studies with this technique there are signs that aged stem cells are rejuvenated whereas young cells lose some regenerative potential.

Old animals with a young partner have, through the shared circulation, continuous access to the young organs. They also are provided with a youthful metabolic environment with frisky pheromones and a lively immune system. In some models there is only an exchange transfusion so the blood in the old mouse is replaced with that of the young one. Astonishingly it was found that the young blood improved old muscle regeneration, and old blood inhibited muscle power and coordination in the youngster. The big question is what is it in the blood that is making the difference. Some reports suggest that restoring youthful levels enhances stem cell and tissue function in the heart, and restores muscle strength and physical endurance.

Several lines of investigation are being pursued. Eventually some sense will come of these findings and, with luck, effective ways to slow ageing and promote rejuvenation will follow.

A chimera is an animal that also has cells from another animal. It sounds pretty creepy but makes sense when you realize that someone who has received a kidney or heart transplant is a sort of chimera. They have a working part of another person within them. Transplants of all sorts are now commonplace and widely accepted. Immature human glial cells have been injected into the brains of mice. They develop into astrocytes and replicate within the mouse brain displacing the mouse equivalent. The human astrocytes (a star-shaped glial – or connecting – cell of the central nervous system) are much bigger than the mouse ones and have up to 100 times more tendrils. Amazingly, on tests for mouse memory and cognition the injected mice are much smarter than normal mice. The same technique offers hope for renewal and repair of myelin sheaths around nerve fibres and thus a potential treatment for multiple sclerosis and other demyelinating diseases.

It does raise the question of whether human brains can be repaired and whether there might even be techniques that would enhance brain function.

25 Extending life: Cryonics and hibernation

"To die, to sleep – to sleep, perchance to dream – ay, there's the rub, for in this sleep of death what dreams may come…"
William Shakespeare (*Hamlet*, Act III, Scene ii)

On November 18, 2016 *The Times* reported on a 14-year-old girl who was the first British child to have her body frozen in the hope of being brought back to life later. In the United States, the Cryonics Institute in Michigan has, at the time of writing, preserved 130 humans and 118 pets with more than 1,000 people signed up to be frozen at the appropriate time. Their website gives the rationale for the treatment. It is headlined: "Envision a Brighter Future". Cryonics gives you a chance at life when medical science has given up. It has the potential to stop or even reverse ageing. You will be reunited with children and grandchildren in the future, and will have the opportunity to witness the incredible things the future has to offer. It promises the elimination of debilitating diseases, an improved quality of life and the possibility of an unlimited lifespan. It also works for animals and can preserve or revive endangered or extinct species, or your pet dog if you think that's more important.

The process relies on draining blood from the body and perfusing the brain with an anti-freeze solution called a crypoprotectant. The body (or just the head if you want to save money) is then cooled over a few days to the temperature of liquid nitrogen, minus 196^{0}C. The expectation is that no metabolic processes will take place at this temperature and the cells will be preserved in a kind of suspended animation. The content of an intact cell is about 70 per cent water. It is an acknowledged challenge to freeze tissues without having harmful ice form in and around the cells.

Vitrification is a method of cooling living cells to cryogenic temperatures without generating ice. Cryoprotectants increase the concentration of solutes in the cells. At very low temperatures the contents of these cells solidify into a glass-like structure without forming ice crystals. Using this method sperm, embryos, blood cells and stem cells are routinely deep frozen, stored for years and successfully thawed. However, conservation of large chunks of tissue and whole organs is technically much more difficult than preserving small collections of cells.

Let us consider the practicalities of this process. First, someone has to die. To do that they will usually be seriously unwell and probably suffer a prolonged period before death with low cardiac output and low blood oxygen values. That won't do their brain and other vital organs any good. Then, after death some sort of initial preservation protocol has to be carried out.

The Cryonics Institute advises immediate cooling with ice after the patient has been declared dead. The body must then be shipped to the Cryonics Institute in Michigan as quickly as possible. They do say that, if feasible, the funeral director should carry out manual or machine ('thumper-type" devices which deliver automatic hands-free chest compressions) cardiopulmonary resuscitation. Is that within the job description of a funeral director, and how long before they start the process? Heparin, a blood anticoagulant, also needs to be given.

If anyone is interested in mechanical chest compression machines I suggest you look at an FDA report from 2013 (see references). Evidence about their use in CPR following a cardiac arrest is insufficient to conclude that they are any good. They may be beneficial or they could be harmful. There just isn't enough information to make a decision. The idea is to keep some blood circulating; enough to prevent brain cells being destroyed before they are deep frozen. There doesn't seem to be any attempt to deliver air or oxygen to the lungs. So even if the blood is sluggishly circulating the process relies on the passive flow of air to provide oxygen. I've heard of flogging a dead horse. These machines are flogging dead hearts. Proponents of cryonics will say, I imagine, that it's better than any alternative.

The protocol for the Alcor Life Extension Foundation is on their website. If staff and equipment are standing by as the patient dies they attempt to keep the circulation and breathing going by attaching a "thumper" machine to the patient immediately after death. If the subject is not in a suitable facility they will have to be moved to one where they are attached to a portable heart-lung machine. In the meantime tubes are inserted into major blood vessels and a mixture of substances infused to stop the blood clotting, to try and prevent cell damage, to maintain blood pressure and to protect the brain. Hooking a body up to a heart lung machine is not that easy and in my experience needs someone who is skilled and practiced. Once connected the machine is used to diffuse cold cryoprotectant through the blood vessels. The next step is to rapidly cool the body to minus 125^0C under fans circulating cold nitrogen gas. The subject is then cooled down further to minus 196^0C over approximately two weeks. Long-term storage is in liquid nitrogen in a thermos flask-type container. If the patient is not in a facility that offers even the most basic elements of this protocol they recommend that a drug is given to stop the blood clotting and the body is packed in ice for prompt shipment to Alcor. Several contentious statements are on the website. One is: "The biological changes known to occur in the first hours following a cardiac arrest are fundamentally minor and reversible in principle" and "technology already exists that could recover patients after more than five minutes of cardiac arrest, although it is seldom used".

During my working life as an intensivist we struggled to keep people alive and used all the resources at our disposal. Maybe there was something we didn't understand, but if we could have kept people alive or brought them back after a cardiac arrest we would have. If we believed that *post mortem* changes were minor and reversible we might have tried even harder. All of us were aware that recovery after prolonged cardiac arrest was possible, but only in a limited set of circumstances, and only if resuscitation was started promptly as soon as the heart stopped. Death cannot be diagnosed for several minutes after the heart has ceased beating and all resuscitation efforts have stopped, and the preservation protocols cannot start until death has been pronounced. This process is nothing to do with artificially prolonging the life of someone who is critically ill. That's what we tried to do all the time in intensive care. It's about what happens when the doctors fail to resuscitate or think that resuscitation will be futile. The patient has died, and has been dead for at least a few minutes beyond the hope of being revived by current medical science.

It will be a challenge, to put it mildly, if there is enough coherent substance left to be restored into a functioning copy of a human being.

There is also a company called KrioRus based near Moscow delivering cryonic services. If you are in the United Kingdom, Cryonics UK is a British-based group of volunteers who have an ambulance and will come and assist with the initial preparation after death before arranging shipment to the chosen cryonics storage provider.

The cryoprotectant is potentially toxic. Too high a concentration and cells are damaged. Too low a concentration and it doesn't work. And does anyone know what the likely effects will be of bathing cells in this solution for years and decades? It takes hours and perhaps days, before the body is ready to be immersed in the liquid nitrogen. We do know that brain cells are exquisitely vulnerable to lack of oxygen. The information on the cryonics websites makes a point that much of the cell damage is caused by reperfusion, restoring the blood supply. They are absolutely correct in this, and they are hoping that in the distant future this will not be a problem. But it is surely wishful thinking to think that cell damage will be slight during the period without any circulation. As cells die the cell wall is disrupted and contents spill out. I would hazard a guess that before vitrification at the cryonics centre an MRI scan of the head of most of the bodies would show a mushy soup where the brain used to be. I'm not aware of any reports of such a scan.

Like Humpty Dumpty there is no way that disrupted cells can be put back together again.

If it does work there is the hope that the extremely low temperature will prevent any chemical reactions taking place and the patient, who let us be reminded had actually died, can be brought back to life and then cured of whatever they died from.

The cost of all this? It seems to vary from about $30,000 to $200,000 for the initial freezing. Will it work? Although there is science and experience behind the basic premise, the actuality relies on a lot of hope. Freezing to minus 196°C and maintaining a healthy body and organs for many years is very different to lowering the temperature to 15° to 20°C for a matter of hours (the temperatures sometimes used in heart and brain surgery). Will medical developments advance to provide a cure? Will the antifreeze solution and process of freezing and reanimation damage or destroy the brain? Will memories and personality remain intact? What sort of world waits on awakening? Not least, will the cryonics company continue to pay the electricity bill and will your future relatives stump up for any future expenses? Who knows? Are deep frozen humans dead or alive?

The dictionary definition of death is permanent cessation of vital functions. If they can be resurrected then logically they were never dead.

There is some scientific work providing limited support for the idea. One study showed that a simple worm could be vitrified and then thawed out and would still react to a smell it had been trained to respond to suggesting that memory had been preserved. The northern wood frog and several other animals can survive semi-frozen without a heartbeat for a number of months. Neurological damage from a period without perfusion may not be universally fatal, at least in cats, and brain tissue isolated from recently dead humans shows some signs of potential viability. If cryonic preservation is going to work it would have its best chance in a healthy animal that is killed quickly, prepared and rapidly deep-frozen. There is no practical obstacle to doing the research to see if this works, even for a matter of days let alone decades.

To date, as far as I am aware, no mammal has been cryopreserved and successfully resuscitated. The answer to any scepticism is simple and incontestable. The techniques have not yet been discovered but they will be in the future. In the meantime those who sign up for this service have to hope that the techniques we have today are good enough to preserve something for those in the future to work with. There's a lot on offer and no comeback if those offering it fail to deliver on their promises. Who will be around to complain?

If all this sounds a bit far-fetched it's worth considering that some animals, including large mammals, can survive for prolonged periods of time without eating, drinking or indeed being active in any way. Torpor is a period of decreased physiological activity, characterized by a reduced temperature and metabolic rate. It allows a broad range of vertebrate and invertebrate animals to put life on hold when times are tough. Many small mammals and birds become torporous every night. Hibernation is an example of torpor lasting weeks or months. Some insects as well as our beloved nematode worm, *C. elegans*, can even stop developing in adverse conditions until things improve. This is known as the "dauer state".

Although hibernation is often associated with low temperatures, this is not essential. Metabolism is drastically reduced to anything from 25 per cent to as low as 2 per cent of normal values, saving up to 90 per cent of the energy that would otherwise be required to maintain temperature. Despite what is said about bears in woods, during many months of hibernation bears do not eat, urinate or defecate. Their heart rate falls from about 60 to 10 beats per minute. Interestingly there don't appear to be any specific genes for hibernation. This suggests that it is caused by the expression of genes that already exist rather than the evolution of new hibernation genes. The critical universal mechanism seems to be reversible protein phosphorylation. This adds or removes a phosphate group from proteins producing major changes in the activity states of many enzymes. Important targets are systems involved in energy supply and consumption. Hibernation is also associated with suppression of DNA transcription and translation, possible through epigenetic changes. Outside of the nucleus of the cell there is global suppression of translational activity mediated through control of messenger RNA. Protein synthesis is also directly inhibited.

A small percentage of genes are actually up regulated and these are of great interest as they are likely to be important in maintaining the animal during hibernation. During this period cold makes the blood viscous and the circulation is sluggish, ideal conditions for blood clots to form. Clotting factors are inhibited and platelet levels reduced making blood clots less likely. Heat shock and similar proteins are expressed to protect against cellular stress. Antioxidant defences are enhanced, possibly because oxidative damage takes a lot of energy to repair, and also to deal with the huge increase in oxygen consumption that occurs when coming out of hibernation.

Hibernators spend up to half their lives in torpor. Does this mean they live longer to catch up on the time lost to activity? The evidence is sketchy but the longest living animals in the rodent family Sciuridae are non-hibernating squirrels. Bats live several times longer than non-flying mammals of about the same size, and hibernating bats appear to live longer than those who don't have periods of torpor. Is this the effect of torpor or of being hidden away from predators for long periods of time? Investigators are not sure.

At the moment there is no way of inducing hibernation in humans. The same basic responses that occur in hibernating animals all occur in humans exposed to various stresses, but are not as intense or implemented as speedily. There appears to be no fundamental reason why the mechanisms used in hibernators can't be turned on in humans.

Perhaps induced human hibernation will become reality in the future and that is how people will wait out the years until medicine finds a solution to their problem.

26 Extending life: Uploading

"I do not fear death. I had been dead for billions and billions of years before I was born, and had not suffered the slightest inconvenience from it."
Mark Twain (no source available)

Cryonics aims to maintain a living brain by putting it into deep freeze. The next logical step is to upload who you are to a computer to dispense completely with the organic bits. This may sound far-fetched but, critics retort, so did heavier than air flying before the early 1900s, or cracking the atom, or building new life through genetic engineering. Perhaps, scanning and mapping all the important functional features of the brain and then copying that to a computational device is the way to make the body redundant. If that's not possible how about replacing individual neurons in an incremental process until the whole brain is non-biological. Some people think that the mind is a result of the dynamic functioning of billions of neurons, and that their actions are governed by physical and electrochemical processes that follow applicable laws. By faithfully mapping a human brain it might be possible to replicate its functions in an artificial memory and calculating device.

In 1936 Alan Turing, of Enigma Code fame, as depicted in the 2014 film *The Imitation Game*, described the principles of a modern computer. The transistor, a key component in computers, was invented in 1947. A big breakthrough in size and computing power came with the invention of the microchip or integrated circuit that put many of the components together on one piece of semiconducting material. A modern circuit may have several billion transistors in an area the size of a human fingernail. Moore's law, named after Gordon Moore (co-founder of the giant electronics company Intel), projects a doubling in the number of components on a circuit every two years or so. This isn't a physical law but an observation of what has happened, and what, so far, is continuing to happen. A 2016 chip has a million times the capacity and a thousand times the speed of a computer of the early 1970s. All of us who use a computer, or a smart phone, or mobile banking, or pay with credit cards are benefitting from these advances. The capacity of computer memory has likewise increased while the cost has plummeted. Copying an individual with all their memories, thoughts, actions and personality to a computer is called mind uploading, mind copying, mind transfer or whole brain emulation. Whatever is on the computer is you, indistinguishable in every way apart from the lack of a body.

This concept is based on the belief that consciousness and identity are somehow contained within the neural pathways in the brain. There are a few examples of people whose brains have been deprived of oxygen for prolonged periods but have woken up with intact minds. One example is Anna Bågenholm. What happened to her is described in more detail in Chapter 7. This is taken as evidence that the mind is not dependent on continuous electrical and chemical activity but can in effect "reboot" as long as it is physically preserved. By extension the mind consists primarily of the physical arrangements of the cells and the molecules in the cells. Much of the metabolic activity is there to keep the cells alive and may not be essential to consciousness, memory and identity. If you accept this, then the arrangements of connections are the most important thing to preserve. The goal of all this would be to make a high quality digital recreation of the brain and simulate it on a computer.

The long-term vision is to preserve the whole brain, scan it non-destructively and then digitise the information. If brains can be preserved whole and scanned without destroying any tissue then some think it might even be possible to restore the original brain and place it in a robotic body or link it with an avatar.

Brain preservation for scientific and medical research has many supporters. It is much more controversial and speculative when brain preservation is promoted as an alternative to death. For those in favour the hope is that automated brain scanning, neuroscience and neural modelling will, in time, make it possible to read preserved human brains well enough to reconstruct limited memories or full self-awareness.

Brain emulation is a potentially interesting and important goal. To do this it is necessary to accurately model the brain and understand how it works. If we could do this it would give insights into the biochemistry and neurochemical behaviour of individual cells and networks. It would undoubtedly have considerable benefits for treating brain problems. The brain is an incredible effective and efficient processing machine. Emulating its processes could lead to better computers with true intelligence. It has the potential for answering important philosophical questions about consciousness, free will and personal identity. Being able to do this would provide a back-up copy of a person, in the same way that information can be backed up on an external hard drive. It is also a way of achieving immortality, living on after the body has died. Although this is far from being a reality there is quite a lot of speculation about how this could be achieved.

The first goal is to preserve a brain without damaging any of its neurons or connections, and ensuring that it doesn't decay. The Brain Preservation Foundation offered a prize for the preservation of an entire small mammal brain. The $26,700 prize was won in 2016 by researchers at 21st Century Medicine (21CM). The company used a combination of ultrafast chemical fixation and cryogenic storage to achieve the near perfect long-term structural preservation of an intact rabbit brain. The next step is to win the prize for preserving a pig's brain.

During the last few years new techniques such as **functional imaging** have started to give neuroscientists better insights into how the brain works. It's long been known that defined areas of the brain control certain functions. The outermost layer of the brain is the cortex which is where thinking and voluntary movement is controlled. The cortex is divided anatomically into several lobes. The occipital lobes at the back of the brain are responsible for vision, the parietal lobes manage sensation and body position, and the temporal lobes are where memory and hearing are located. Brain control of muscle activity is relatively easy to study. Electrical stimulation of specific areas in what is called the motor cortex causes specific movements to happen and this has been accurately mapped to corresponding bodily movements. It is more challenging to work out where thinking occurs, emotions are generated and memories are made and stored. But progress is being made.

To make progress much more detail is needed about individual neural pathways and how they work. The Human Connectome project (**www.humanconnectomeproject.org**) is attempting to map human brain circuitry comprehensively. The idea is to marry up information about connectivity with behaviour, and to identity the impact of genetic and environmental factors on these findings. Anatomy and function are investigated with magnetoencephalography (MEG), electroencephalography (EEG), and several types of different MRI scan. The MRIs record the anatomy of the brain, look at the blood flow in resting states and when specific tasks are carried out, and focuses down on the white matter to examine neuronal connections. The Gallery (**www.humanconnectomeproject.org/gallery/**) has some stunning images of the human nervous system. Even Facebook has become involved in this endeavour by hiring a team of scientists to work on brain computer interfaces.

The Russian billionaire, Dmitry Itskov, is the founder of the 2045 Initiative. Its main scientific aim is to create technologies enabling an individual's personality to be transferred out of the body thus extending life, perhaps to immortality. The target date is the year 2045 (unsurprisingly given the name of the initiative). The website (**http://2045.com**) shows the milestones leading up to this date. They start with a robotic remotely controlled avatar, progress to an avatar into which a human brain is transplanted and then go on to an avatar with an artificial brain to which the human personality is transferred on death. It culminates in 2045 with the ultimate aim of a substance-independent mind with a body that can take any form and be geographically independent.

It will be a major challenge to accurate map and model the human brain. In the brain of a young adult there are about 1,000 trillion connections (10^{15} connections). It is not only the number of connections that is important but also what happens at the synapses. There is a range of different synapses with a variety of transmitters leading to an enormous number of ways that nerve signals can combine, inhibit, enhance and modify. Brain activity is also plastic, changing with learning, activity and memory. To be able to process all the information computers would have to be much more powerful than current machines. In 2013 a Fujitsu K supercomputer simulated one second of neuronal activity. It took 40 minutes for the simulation to run using 82,944 processors and one petabyte (one quadrillion bytes) of main memory. The next generation of computers will be 100 times faster but there's still a way to go before they are anywhere near big enough to be able to upload a human brain. Even if supercomputers capable of the task existed their power consumption and heat production would be enormous, much greater than the 25 watts the human brain generates. If you think that this problem is just too big and complicated, commentators remind us that simulating complex systems such as individual proteins or genomes would have seemed unachievable not so long ago.

Eventually computers may have the capacity to contain a human brain but we're a long way from knowing what to upload.

At best, however, uploading would create a copy and not the original person. The brain almost certainly contains chaotic dynamics suggesting that the workings of the mind are not completely deterministic solely based on chemicals and connections. If this is the case, no matter how accurate the emulation, it is likely to deviate from the original over time. The brain may also have quantum states that are impossible to measure. If this is true even if you could upload my brain the electronic version will develop and evolve into something different from me, a close cousin, a wicked sibling or perhaps my wiser, kinder, better half. Surely, it wouldn't take much effort to refine the model to discard the bad and keep what's good.

We exist within a corporeal body, are sustained by it, informed through its sensors, respond through its actions and are influenced by its interactions with the real world. Without a body I would not feel hunger or pain. There would be no circulating hormones to respond to fear or pleasure. Would it be possible to feel joy or love or anger? How much of the body would need to be virtually created to complete the emulation?

Can an individual's consciousness be in more than one place? How can we tell true consciousness from that of a mindless computerised symbol manipulator? These are the type of questions to keep philosophers happy for years. If mind uploading can be achieved they may be able to stop their speculation and find the answers. There are also a whole heap of ethical and legal questions that would arise. Would the emulation have legal rights? Could it take out a mortgage, own property, sue for slander? What would happen if it commits a crime, could it be taken over by computer malware, would biological humans feel threatened? The more I think about it the more it boggles my mind. But most of these are relevant questions. Not so much because mind uploading is imminent, but because intelligent and conscious computers are a real possibility in the not too distant future.

The idea of emulation may seem far-fetched but it is receiving attention from legitimate researchers. It may never come about but in the meantime it is stimulating ideas and experiments that will help improve our understanding of what it is to be human and what it may take to live for hundreds of years.

27 Extending life: Head transplant

"Life asked death, Why do people love me but hate you? Death responded, Because you are a beautiful lie and I am a painful truth. **Anon**

In the 1983 film *The Man With Two Brains*, Steve Martin falls in love with a brain kept in a liquid-filled jar. Other films and books feature disembodied brains. They make great stories but we've a long way to go before a brain can exist as an isolated organ on life support. The medical name for the procedure of sticking a head from one animal onto the body of another is *cephalosomatic anastomosis*. Simply merging the Greek words for head and body together doesn't give it legitimacy, but by the time you read this a human-to-human head transplant may already have taken place.

At a basic level the body can be considered to be a life support system for the brain. There are many people alive who exist functionally as a head attached to a body over which they have no control. These are people who have had a high level disruption of the spinal cord in the neck. The spinal cord is the bundle of nerves providing the connection between the brain and the rest of the body. It carries all sensation up from the body to the brain, and sends instructions from the brain down to the body. Tetraplegics or quadriplegics – both words are used – have lost the use of all four limbs. The most important muscle used for breathing is the diaphragm. Fortunately for some tetraplegics the nerve supply for the diaphragm, the phrenic nerve, comes from the middle of the neck. If the spinal cord is disrupted below the origin of the phrenic nerve the patient can often breath for themselves. Otherwise they will depend on a machine (a ventilator) to pump air into their lungs.

The muscles in the human body are of two types. They are either voluntary that we can control ourselves. Examples include leg and arm muscles, and also the muscles we use for breathing. These all need an intact spinal cord. Or they are involuntary muscles over which we have no conscious control. These include the heart and most of the muscles in our gut as well as those that dilate and constrict the pupils of our eyes. These muscles can continue to work even if the spinal cord is severed. Although breathing depends on voluntary muscles we don't have to consciously decide to take each breath. There are breathing centres in the brain that respond to high values of carbon dioxide or low values of oxygen in the blood. Try to hold your breath and it is normally the increase in CO_2 that will make you breathe again. If you are on the top of Everest without oxygen it is the hypoxia that will drive your respiration.

A tetraplegic person is, therefore, not able to move, although their heart will carry on beating. The body and its organs, such as the kidneys and liver, will continue to exist sustaining the body and providing energy to the brain, and removing waste products. However, without intensive 24 hours a day maintenance their bodies will not survive long. Unless the body is turned regularly and frequently the skin will break down from pressure sores, nutrition has to be given by tube feeding, and bladder and bowel need assistance to evacuate. Even if the diaphragm works chest infections are common and, if the patient relies on a ventilator, chest infections are inevitable. Maintaining a functioning body in these patients is a daily battle. Although many tetraplegics live long and fulfilling lives, they are likely to be considerably shorter than able-bodied people.

So why are some people seriously considering performing a head transplant, although this Frankenstinian suggestion should really be called a body transplant? One person who has generated a lot of publicity with this idea is the Italian neurosurgeon Sergio Canavero. It is true that the occasional person ends up in an intensive care unit with a permanently destroyed brain but a perfectly healthy body suitable for donation. These unfortunate people are already the major source of other organs for transplantation. There are also people whose bodies will, in time, not be capable of keeping their owner's brain alive. These include those with degenerative muscle or nerve disease, or where cancer is ravaging the body. The idea, therefore, is to marry a healthy body with a dead brain, with a healthy brain supported by a diseased and dying body.

The idea is not new. The first dog head transplantation was actually carried out in 1908. Alexis Carrel and Charles Guthrie who carried out this procedure were not charlatans. Carrel was awarded the Nobel Prize in Medicine in 1912. According to some sources

Carrel was no more than an assistant to Guthrie but Guthrie's head transplant experiments prevented him being awarded the prize. In 1954 Dr Vladimir Demikhov of the Soviet Union created some grotesque two headed dogs. Most died of acute rejection within a few days. The longest survived for 29 days. If you can bear to look, videos of them are available on YouTube. A review committee of the Soviet Ministry of Health decided that his work was unethical but reports say he carried on regardless.

In 1965 an American neurosurgeon, Robert White, grafted six dog brains onto recipient dogs and in 1970 he transferred the heads of four monkeys to different monkey bodies. Within a few hours of surgery the heads could chew, swallow food, bite and track objects with their eyes. These creatures survived for between 6 and 36 hours. More recently in 2015, a Chinese surgeon, Xiao-Ping Ren, described a refined surgical technique in mice. Half his mice survived for more than 24 hours with the longest survivor being six months. Canavero has now managed to transplant the head and shoulders of one rat onto the body of another.

There are not many structures that pass through the neck and have to be joined to attach the head to the body. The blood supply to the head is all through only four arteries, a carotid artery and vertebral artery on each side of the neck. The blood returns to the body through jugular veins, one on each side.

The big challenge is keeping the recipient's brain alive during surgery. If the blood flow is interrupted for more than a minute or two, brain damage occurs.

There are several techniques for protecting the brain and extending the time for which it can be deprived of oxygen. Extreme cooling is one of these methods and is used commonly in heart surgery in children when the circulation has to be stopped for some time. Canavero envisages cooling the head down to 12^0 to 15^0C from the normal body temperature of 37^0. Another possible technique would be to maintain circulation to the head on an artificial machine similar to those used in heart surgery. Steps must be taken to ensure that blood does not clot blocking the circulation.

The tissues around the neck are cut and the major vessels on the head joined with tubing to the corresponding vessels on the donor body. Once the arteries and veins have been joined circulation can be restored and fingers kept crossed in the hope that the brain has not been permanently damaged. The surgeon can take his time joining the other structures. They are the oesophagus (food pipe), trachea (airway), bony spine, muscles and skin. All these structures have the potential to join and to heal, although none of this is without technical difficulties or risks. Canavero also has a protocol for reattaching the spinal cord, as well as other more minor nerves such as the phrenic nerve. It is his plan to sever the spinal cord cleanly with an extremely sharp knife. It is his hope that a clean and fresh cut will increase the chances of being able to join the spinal cord and get it working again. Unless the nerve supply can be restored the head/ body combination will be the same as a tetraplegic with the added disadvantage of having to fight rejection.

Unfortunately nerve tissue does not regenerate, or at least that was the belief until not so long ago. Encouragingly, not least for tetraplegics and others with spinal cord injury, there is early evidence that nerve cell regrowth may be possible. Chemicals such as polyethylene glycol are able to fuse together severed axons or injured neurons. Collagen conduits containing a growth promoting gel may also assist neuronal regeneration, and electrical stimulation may accelerate regrowth. Stem cells or olfactory ensheathing cells may also have a role to play with their ability to be transformed into new nervous tissue. This all provides some hope that the nerve function can be restored. At the time of writing all we have is anecdotal evidence that this might be possible.

Rejection is another major issue that needs to be addressed. Much has been learned from other organ transplants and immunosuppressive agents will have to be given to prevent rejection. The highly immunogenic nature of skin was only overcome as recently as 1999 with a new combination of immunosuppressants. Since then hand and face transplants have been performed. Whether these drugs will be effective in a head transplant is unknown. Unlike a rejected kidney, a rejected head cannot simply be removed and another one stuck on.

It might be technically possible surgically to take a head from one body and put it on another. It's not something I would personally contemplate volunteering my head for, at least until it becomes possible to form a functioning neurological connection between the head and body. There will be substantial issues around rejection. Even if it is feasible the procedure will be expensive, rehabilitation will be prolonged and psychological problems may be overwhelming. It hasn't happened yet, and maybe it never will.

The ethical and practical implications are considerable and the technique is likely to find more use as a plot device in a science fiction movie than in a practical technique for extending life. If a human-to-human transplant works, how much more unacceptable would it be to mount human heads on other animals? The mythical centaur had the upper body of a human and the lower of a horse. What body would you choose – a pig, a dog, a giraffe? The implications are really mind-boggling. At the time of writing Canavero claims that his team have already transplanted a head from one cadaver to another. It is said that he will attempt it on a live patient sometime in 2018.

Other, possibly less gruesome, approaches are also being investigated. Robot technology is developing all the time. Machines have been made that can walk and run, work machinery, and act as eyes and ears, and even touch, for a human controller. It is a small step from control being exercised through a keyboard or touch screen to having thoughts relay instructions. Imagine being trapped within a useless body but being able to wander far and wide through a thought controlled robot? If you've seen the film *Avatar* you'll know what I'm on about. But instead of an organic flesh and blood avatar, you are in control of a metal and plastic extension of yourself. These techniques are likely to be possible and they would have wide applications. Already surgeons are able to operate remotely using robots. There is no technical reason why a doctor based in an office on one continent can't operate on a patient thousands of miles away. Engineers can operate machinery in hostile environments such as in nuclear power stations or at the bottom of the sea. Couple virtual reality headsets to remote inputs and exploration could take place from the comfort of an armchair.

Although this technology may not extend the duration of life it will allow those who are physically impaired to participate more fully in the world about us, and that has got to be a good thing hasn't it?

And instead of relying on a human body how about mounting the head, or just the brain, on a robot. Many of the functions of an organic body can be dispensed with. There would be no risk of pressure sores or having sore joints. The brain would need some glucose for metabolism but not much else. With batteries to power most functions food becomes unnecessary, and so does the gut, the liver and kidneys. I suppose you might still get a headache, but most other illnesses would be impossible.

28 A world of millenarians

"If things seem under control you are just not going fast enough."
Mario Andretti (Italian-American racing driver)

Some physicists believe in the multiverse, a countless number of parallel universes constantly spinning off into time and space. If this reflects reality then millenarians may well already be living in a duplicate London wondering on the short brutish 80 year-old lives lived by previous generations. Physicists also seemed to be puzzled by the nature of time itself. We perceive it as a linear continuum, one minute following another, always going forward. Perhaps we can change time and long life will be more to do with the way the clock ticks than any fundamental change in the human body. Then perhaps we are simply figments of the imagination of someone else, living in a cosmic computer game, mental constructs fated to be rebooted when Ctrl-Alt-Delete is pressed. I'm going ahead on the basis that we all, more or less, share the same reality. That we are born, grow old and then die.

This gives me the opportunity to speculate a bit about what the world might look like if the period between birth and death becomes many hundred of years longer than is possible today.

I'm going to assume that the average lifespan can be extended to 1,000 years or thereabouts. Let's hope that for much of that time people are going to remain physically and mentally active and in good health. If not, we are going to need many more nursing homes, and carers, and incontinent supplies. Furthermore let us assume that the techniques for prolonging life are widely available and not just restricted to a small group with the cash and connections to buy longevity. The rich and powerful may want to keep immortality for themselves, but I'd prefer to ponder a world where the potential benefits are accessible to all. This is an opportunity to give my imagination free reign because, inevitably, anything I come up with is highly speculative, and only those who touch immortality will be able to say whether it comes to pass.

Before I get too carried away it may be worth injecting a few cautionary words. Even without direct genetic manipulation we, as a species, may already be doing harm to our genome. Compared with other common species humans have low genetic variation, in the order of 0.1 per cent to 0.2 per cent, reflecting our common descent from a small founding population. This may increase our vulnerability as a species. In addition, civilisation and medical advances have diminished the pressures of natural selection. Each of us has some 100 new mutations not present in our parents, of which a few are likely to be harmful. In the past, genetic weaknesses were often removed as people died young. Nowadays reproductive potential and longevity are maximised without regard to genetic issues. One possible consequence of these actions is a long-term decline for our species in overall physical and mental abilities.

Thomas Robert Malthus (1766-1834) is famous for writing a book on population growth, *An Essay on the Principle of Population*, published in 1798. He raised the spectre that an expanding population would outgrow its resources such as food supply, water, living space and so on. It's a familiar argument because the issue, far from going away, is very much a concern today. It's salutary to realise that Malthus was worried as the world's population neared one billion. It had taken the whole of human history to reach that figure.

H. sapiens probably evolved about 200,000 years ago. The Neolithic Era was about 12,000 years ago and coincided with the development of farming and thus early civilisation; not so long ago in historical terms, or in the context of 1,000-year-old people. During those thousands of years civilisations rose and fell, whole communities were found to be unviable and, at times, the existence of the human race must have seemed precarious and threatened.

Even in more recent historical times plagues such as the Black Death wiped out one-third of the people living in Europe. It took until around 1800 for the human population to reach that first billion. It took a further 123 years for the population to double to two billion. The next billion took a mere 33 years, the billion after that 14 years, and each subsequent billion has taken around 13 years. Today's world population is estimated to be about 7.5 billion. Despite the fears of Malthus and many others, the world is producing more food than ever before, and almost everywhere people are living longer and healthier lives. It is human conflict, corruption and lack of cooperation, not resources that are the main factors in preventing access to safe water, adequate food and essential public health measures.

Malthus will almost certainly be proved right at some time. The world's population cannot go on expanding indefinitely in a world of finite resources. Some commentators believe that there are signs that population growth is slowing down and could reverse towards the end of this century after a maximum population of 10 to 12 billion is reached. Smaller families are now the norm in wealthier countries. When having only one or two children still gives you a good chance of passing on your genes the pressure for large families lessens. Modern contraception has made getting pregnant a choice and not an uncontrollable side effect of sex. And smaller families are associated with a higher economic standard of living, better education and more leisure time.

Developments including modern farming techniques, genetically modified crops and hydroponics have the potential to further increase the world's food output. If cheap, non-polluting energy sources become widely available much more could be done to provide clean water, heat and light. Most of all knowledge can be disseminated with access to the Internet, and knowledge is probably the best way of helping people work towards a better future for themselves and their descendants. Who knows, mining the asteroids may become a reality and energy could be beamed down from space. There is still plenty of living space here on earth if the oceans and subsurface areas are to be exploited. But there is almost unlimited potential freely available on the Moon, Mars and the rest of the known universe. What would a few years or a decade or two matter to astronauts who anticipated living for many centuries?

Let us start with a bit of simple mathematics. We have seen earlier how the single cell from which we all come multiplies into billions of cells over a short time. If we do a similar calculation we can easily see what effect extreme longevity will have on population. It is reasonable to assume that a mother can have a child at any age she chooses. We have already had 60-year-old mothers. Even now eggs can be frozen for later use. It is entirely feasible that somatic cells can be turned back into eggs (or sperm), and even that a father's cells could be enticed to become an egg. As a side effect, sex will not be necessary for reproduction and may, in fact, be discouraged. Wouldn't it be better, and considerably less messy, to send off a few skin cells from mother and father (or indeed father and father, or mother and mother) to be made into embryos? Those grown can then be screened for inherited diseases and unwanted traits or behaviours. The future offspring is then carefully selected for implantation. For mothers too posh to push there will be surrogates or artificial wombs, ready carry the foetus during those annoying and unpleasant months until delivery.

Let's assume a parent (a mother or a father) decides to have a single child at 50. When that child is 50 they have a solitary offspring who in turn has a child at age 50. You get the idea, I hope. In this scenario when our millenarian has their 1,000th birthday party they will have 20 direct descendants, children, grandchildren, great grandchildren etc., to be at their party. However, instead of having one child every 50 years the parent has two, who themselves have two each and so on, there will be an extraordinary 2,097,150 offspring, twice the population of Birmingham, to be invited and a much bigger hall will be required. If they each have 3 children fifty years apart, the number of direct descendants (not including spouses, second and subsequent marriages etc.) will be 5,230,176,600 and a new planet would be called for. The table below may make things clearer.

Number of direct descendants from 2 children every 50 years

50 years	2 children	2
100 years	2 children each have 2 = 4	2+4 = 6
150 years	4 each have 2 = 8	6+8 = 14
200 years	8 each have 2 = 16	14+16 = 30
250 years	16 each have 2 = 32	30+32 = 62
300 years	32 each have 2 = 64	62+64 = 126
350 years	64 each have 2 = 128	126+128 = 254
400 years	128 each have 2 = 256	254+256 = 510
450 years	256 each have 2 = 512	1022+512 = 1022
500 years	512 each have 2 = 1024	1022+1024 = 2046

and so on . . .

Longevity

The 1,000th birthday party for our first millenarian will be fascinating. Of course, mundane problems such as baldness will be cured during the 21st century, sadly too late for me. Whether or not to have bodily hair will be a lifestyle decision. Genetic manipulation will be routine. You may choose to spend a couple of hundred years as a female before trying the masculine form. Some futurologists say we will be immobile obese beings with withered limbs transported around on levitating platforms. My view is much more optimistic. For starters, it will be possible to eat what you like without getting fat. It can't be an insuperable task to manufacture non-fattening chocolate or make a minor genetic manipulation to increase metabolism or prevent fat being laid down.

Anyway with more leisure time people will want to keep their bodies healthy. I predict there will be plenty of gain for little or no pain. Instead of personal trainers standing by putting you through exercises, our descendants will simple loan the trainer their body to inhabit for the workout, while the rest of them stays at home and eats cake. Transgenic engineering will be common. Grow some gills, fluoresce in the dark, have fur? Why not?

Even now we are on the threshold of being able to take a constant stream of photographs to record our days. Surely every moment of future lives will be recorded in HD 3D smell, taste and touch detail. If you want to know where you were and what you were doing at any point in your life it will be easy to recall. With all that data uploaded, the difference between material and extracorporeal life would blur. In fact, bodies will probably be optional. At the party there would be people of all ages from new born to those in their nine hundreds. Those off-world or in hibernation would be represented by their avatars. The interface between man and machine would be seamless. Exoskeletons will let humans lift huge weights and run at the speed of present day cars. Pure thought will control machines and allow silent communication between individuals, telepathy made a reality.

Merely think of a question and a quiet voice in your ear will answer it. Who was the 45th President of the United States? The voice will whisper President Trump. What was the United States? A short-lived democracy that flourished in the 20th Century. What is a

democracy? A representative form of government that died out after direct citizen decision-making became established – see Brexit, and so on.

Much will depend on how these new superhumans age. It is easy to conjure the appalling spectacle of old age homes filled with gaga 900-year-olds with an expectation that they will go on for a further 100 years or more. It will probably be important what treatment is necessary to keep these people alive. During my working life as an intensive care doctor about a third of the really sick patients we cared for did not survive their hospital treatment. When treatment was futile and was only prolonging an inevitable end, it was common to either withhold or even withdraw treatment. This is very different from an active decision and intervention to end a life. If continual treatment is needed to maintain life then it may be possible to simply withhold therapy from those thought unable to benefit. It would be a small step beyond that to cut off the supply of drugs or treatment for those who have reached a certain age, or can no longer pay, or society in some other way decides are not worthy of support. This minefield will be for others to tread through, it's too dangerous for me. As of June 2016 euthanasia was legal for certain conditions in four countries (Netherlands, Belgium, Colombia and Luxembourg), and assisted suicide in four countries and some States in the USA.

Some people already find our short lives too much to contemplate. For others an extended lifetime would be a curse, not a blessing. I suspect that death will be a managed process not the end result of an untreatable disease.

It would be wonderful to envisage a flowering of arts and sciences with people freed of the need to work to live, or to spend much of their life caring for children, but able to devote centuries to study and to self improvement. In 1993 Anders Ericsson wrote an often misquoted paper. This exhaustive review and observational study looked at the relationship between performance and practice focusing on students learning to play the violin. The somewhat distorted message that has been disseminated is that 10,000 hours of practice is needed to become an expert in almost anything (Google the 10,000 hour rule; but also see the post by K Anders Ericsson. *The Danger of Delegating Education to Journalists* **https://radicalscholarship.wordpress.com/2014/11/03/guest-post-the-danger-of-delegating-education-to-journalists-k-anders-ericsson/**). That equates to some 20 hours a week for 10 years. Clearly this is an over-simplistic interpretation of the evidence. I'm convinced that no matter how much I practice and how great my motivation I would never be able to produce anything other than a strangled shriek from a violin. Golf courses are populated by players who try their hardest week after week without any discernible improvement. Nonetheless the core point is generally accepted; if you want to get good at something you have to practice. Innate talent and motivation are important but you have to also put in the hours.

With hundreds of years of healthy life to be filled let us speculate about what can be achieved. Dedicating 10 or 20 years of your life to become an expert in something does not seem extravagant, and you then have plenty of time to become an expert in something else. A mere three years in higher education would only allow you to dip your toe into a topic. A PhD would probably become the starting point for any sensible attempt to become involved in an academic subject. With sport and other physical activities much would depend on the physiological age of our bodies. Most elite sportsmen and women are over the hill by their mid-thirties. Let's hope attempts at increasing longevity manage to freeze degeneration at a suitable age. If I still had the body of a man of 30 and 10,000 hours of practice to work on my skills I could harbour ambitions of playing striker for the England football team. Sadly if they manage to keep me young they can do the same for Harry Kane, Jamie Vardy and Wayne Rooney, which would blow my chances. Perhaps they can even clone Stanley Matthews, Bobby Charlton and Jimmy Greaves. England would finally have a settled football squad, at least for a few hundred years.

Today one in three marriages ends in divorce suggesting that a mere 20 or 30 years together is more than many couples can handle. If 50 years together is worth being called a golden anniversary, who knows what 900 years together would signify. I can't see how the conventional concept of marriage for life could withstand the strain.

Perhaps limited contracts would be the norm and spouses would be exchanged every few years in just the same way a leased car is traded in for a new one. Nowadays, a sugar daddy may be 20 or 30 years older than the young women he is going out with. Such a disparity in age would be a trifle, particularly if ageing is suspended. What would it feel like going out with someone hundreds of years older or younger? Try to imagine your partner or spouse being born in the Middle Ages with the accumulated history and experience that entails. These days it only takes a few years for musical tastes and fashions to change. Would they regret the passing of madrigals and the birth of rock and roll? I'm still stuck in the 60s with Eric Clapton, The Stones and Meatloaf as my idols. Luckily their music lives on but other generations have their own heroes.

People are already having to work longer and take their retirement later. With the expectation of 900 years or more of productive life, patterns of work and leisure will have to change. There will be time to have several careers and become expert in more than one walk of life. Want to learn a language; no problem. Five years to become fluent will be a minor diversion. Just like actor Bill Murray in the film *Groundhog Day* why not start learning the piano in later life? You will have many years to dedicate to learning the skill. With all this time and good health there could be a renaissance in the arts and sciences. It could become routine to dedicate a century or two to good works, to study and contemplation. On the other hand, boredom could find an outlet in social unrest and mindless computer games. Even today technology can leave the old behind unable to use a computer or programme the television. Will people be able to adapt in a dynamic changing world. Perhaps change will slow down or stop as those in power hang on stuck with the attitudes, culture and technology that they grew up with.

I think boredom is likely to be a big issue in our millenarian future. With hundreds of years stretching endlessly ahead keeping occupied will be a challenge. Mechanisation is likely to automate many occupations from factory worker, to pilot, to driver and even sex worker. Almost anything you can think of can probably be done by a machine, more quickly and more effectively than a flesh and blood human. How would it make people feel to be second rate to a robot? It is easy to see how the population could be infantalised and without purpose, expecting to be told what to do and to have their needs satisfied without having to make any effort. In Western democracies we live in an era where individual autonomy is often said to be paramount. Would it be able to survive with competition for resources and with the younger generations clamouring for those in positions of authority to move aside and let others have a turn? Would people have to work, and for how long to gain a pension? Would productive people be willing to pay taxes to support those who had decided to while away the centuries sat on their backsides watching daytime TV? What jail deterrence could you give to someone with an enormously extended life expectancy? And how much more of a crime is it to end the life of someone with so many more years to be lived?

The United States Declaration of Independence states: "We hold these truths to be self-evident, that all men are created equal, that they are endowed by their Creator with certain unalienable Rights, that among these are Life, Liberty and the pursuit of Happiness." Although men (and women) shall have the right to pursue happiness, actually achieving it can be elusive. As the world has become wealthier basic needs are no longer a concern for most people, at least in the wealthier West. Until recently getting sufficient food, warmth and shelter to survive the day was enough to keep an individual occupied. As times have got easier not only has recreational shopping been invented, along with stiletto heels and handbags, but so has train spotting, stamp collecting, bowls and cricket.

Material wealth is a goal for many, but can lead to envy and dissatisfaction. Having money, health and a long life is no barrier to being on antidepressants or committing suicide.

Even today life is long enough so that I can contemplate spending the time to write this book. Think what writing could be inflicted upon the world with millions of multi-centenarians with wide life experience and the desire to share it with their offspring and their friends. If only my immediate family buy this book I could generate a handful of sales. With hundreds of thousands of close family I would have a good chance of getting in the best sellers' list and providing a suitable Christmas present for all at the same time.

I also wonder what guaranteed longevity would do to attitudes to risk. It's a bit of a conundrum that the young, with most to lose, seem willing to take more risks. It appears that suicide bomber is a career (a very short career) for the young. I would have thought that the old have much less to lose. When a parachuting accident or racing car crash can shorten a life by hundreds of years would the young be risk adverse? What would people think about religion? With the afterlife postponed for so long how relevant would God seem? If we can contemplate a future where genes are manipulated to increase longevity perhaps a few small tweaks of the genome will also be necessary to weed out competitiveness, greed and envy.

In the same way that a washing machine comes with a guarantee, perhaps extra centuries of life would have to come with some anti-aggression genetic manipulation.

Genghis Khan took fewer than 20 years to conquer the known world before dying at the age of 65. Stalin was leader of the Soviet Union for 28 years and during those years is said to have been responsible for the deaths of 20 million people. Robert Mugabe became President of Zimbabwe in 1987 and managed to transform his country from being the breadbasket of Africa to the basket case of the continent before being forced to resign aged 93 in November 2017. History is replete with other examples of individuals who have amassed and clung onto power and have inflicted misery upon their subjects.

Queen Elizabeth II was born in 1926 and has been queen since 1952. Her heir, Prince Charles, is in his late 60s and is still not King. Charles' grandson, George, was born in 2013. If he eventually becomes King, he may reign for centuries leaving generations of princes (and princesses, since the law of succession was changed in 2013) waiting in vain to take their place in the line to inherit the crown. Having Nelson Mandela as President for a couple of centuries may be acceptable or even welcome, but a 500-year rule by Mugabe or Kim Jong-un sends shivers up my spine. Many are grateful that any reign is limited in years until the faculties of the leader fades and eventually death intervenes. The knowledge that someone could hold onto power for hundreds of years is frankly scary. Even in benign democracies would our politicians be satisfied with ten or twenty years at the top when they could expect to have several hundred years of productive life?

Wealth and power often go together. Some individuals have amassed extraordinary fortunes in their lifetimes. Society can usually rely on the second or third generation to blow the money or see it distributed or dwindle away through taxation and inheritance. Income inequality is an important issue with some one per cent of adults owning 45 per cent of the world's wealth. A 2016 report from Oxfam stated that the eight wealthiest individuals (yes I really do mean eight) in the world have a combined wealth equal to the bottom 50 per cent of the population. That does seem far fetched. However, apparently in the United Kingdom the top 3,000 earners pay more tax than the bottom nine million, and this is despite the efforts the wealthiest are said to take to avoid paying tax. A world where power, money and status is accumulated and hoarded for hundreds of years could make today's distribution of wealth look equitable. It would be nice to think that systems would be put in place to prevent the concentration of power and wealth in a few hands. I wouldn't bet on it though.

Our species, *H. sapiens*, developed small genetic changes many thousands of years ago that differentiated us from other Hominines. When it becomes possible to reliably manipulate our genome how big a change will be necessary before a different species is created? Perhaps there will rapidly arise what is, in effect, a range of species based loosely on our parent genome? Today we communicate instantaneously over thousands of miles, we cross the skies in huge metal flying machines, we understand and cure diseases that were once thought to be caused by evil spirits. If our ancestors of 1,000 years ago could see us now they might consider us Gods. Our near immortal descendants will be able to manipulate and create life. Perhaps that's as close to God as it is possible to become.

References

Books
Bertil Marklund. *The Nordic guide to Living 10 years longer*. English translation published by Piatkus 2017. Evidence-based guide to remaining healthy.

Mark O'Connell. *To be a machine*. Published by Granta 2017. An exploration of transhumanism.

Richard Dawkins and **Tan Wong**. *The ancestor's tale*. Published by Weidenfeld & Nicholson 2016. The story of the genetic descent of man.

Siddhartha Mukherjee. *The gene: An intimate history*. Published by Bodley Head 2016. A description of the quest to decipher the genome.

Susan Greenfield. *A day in the life of the brain*. Published by Allen Lane 2016. The mystery of consciousness.

Allen J.Tobin and **Jennie Dusheck**. *Asking about life*. Second edition published by Harcourt College 2001. A comprehensive review of life starting with the chemistry and cell biology, encompassing basic genetics and evolution and including an overview of physiology.

Dan Buettner. *The Blue Zones*. National Geographic 2010. Lessons from the zones where people live longer.

How the brain works. *New Scientist* series. Published by John Murray Learning 2017. Multi-author essays about the human brain.

Introduction
Bulterijs S *et al*. It is time to classify biological aging as a disease. Frontiers in Genetics 2015:6;Article 205.

Chapters
1. Shaping our own destiny
Lykouras E *et al*. Searching the seat of the soul in Ancient Greek and Byzantine medical literature. Acta Cardiol 2010:65;619-26.
Mehta P & Dhapte V. Cupping therapy: a prudent remedy for a plethora of medical ailments. Journal of Traditional and Complementary Medicine 2015:5;127-34.
Spurious correlations: www.tylervigen.com/spurious-correlations

3. DNA: The book of life
Cedikova M *et al*. Multiple roles of mitochondria in aging processes. Physiol Res

2016:65 (suppl 5);S519-S531.
Chen BH *et al*. DNA methylation-based measures of biological age: meta-analysis predicting time to death. Aging 2016:8;1844-59.
Hodgkin J. What does a worm want with 20,000 genes? Genome Biology 2001;2:1-4.
Johnson W *et al*. Beyond heritability: twin studies in behavioral research. Curr Dir Psychol Sci 2010:18;217-20.

4. Cellular life
Bianconi E *et al*. An estimation of the number of cells in the human body. Ann Hum Biol 2013:40;463-71.
Spalding KL *et al*. Retrospective birth dating of cells in humans. Cell 2005:122;133-43.

5. Growth
Bhardwaj RD *et al*. Neocortical neurogenesis in humans is restricted to development. PNAS 2006:103;12564-8.
Bliss TVP *et al*. Long-term potentiation. Phil Trans R Soc 2003:358;607-11.
Bliss TVP *et al*. Synaptic plasticity in health and disease: introduction and overview. Phil Trans R Soc B 2013:369;20130129.
Christopoulos GI *et al*. Neural correlates of value, risk, and risk aversion contributing to decision making under risk. J Neurosci 2009:29;12574-83.
Debowska W *et al*. Functional and structural neuroplasticity induced by short-term tactile training based on braille reading. Frontiers in Neuroscience 2016:10;Article 460.
Fairchild G *et al*. Cortical thickness, surface area, and folding alterations in male youths with conduct disorder and varying levels of callous-unemotional traits. Neuroimage Clin 2015:8;253-60.
Gabi M *et al*. No relative expansion of the number of prefrontal neurons in primate and human evolution. PNAS 2016:113;9617-22.
Herculano-Houzel S. The remarkable, yet not extraordinary, human brain as a scaled-up primate brain and its associated cost. PNAS 2012:109;10661-8.
Kandel ER. The molecular biology of memory storage: A dialog between genes and synapses. Nobel Lecture 2000. www.nobelprize.org/nobel_prizes/medicine/laureates/2000/kandel-lecture.pdf
Lynch MA. Long-term potentiation and memory. Physiol Rev 2004:84;87-136.
Olkowicz S *et al*. Birds have primate-like numbers of neurons in the forebrain. PNAS 2016:113;7255-60.
Tobler PN *et al*. Risk-dependent reward value signal in human prefrontal cortex. Proc Natl Acad Sci U S A 2009:106;7185-90.

8. Life expectancy
Barbieri M *et al*. Data resource profile: The human mortality database (HMD). International Journal of Epidemiology 2015:1549-1556.

Burger O *et al*. Human mortality improvement in evolutionary context. PNAS 2012:109;18210-4.
Chamberlain G. British maternal mortality in the 19th and early 20th centuries. J R Soc Med 2006:99;559-63.
Christensen K *et al*. Ageing populations: the challenges ahead. Lancet 2009:374;1196-1208.
Colchero F *et al*. The emergence of longevous populations. PNAS 2016:E7681-E7690.
Rowbotham J and Clayton P. An unsuitable and degraded diet? Part three: Victorian consumption patterns and their health benefits. J R Soc Med 2008;101:454-62.
Statistical bulletin: Life expectancy at birth and at age 65 by local areas in England and Wales: 2012 to 2014. Office for National Statistics, release date 4th November 2015.
Statistical bulletin: National life tables, United Kingdom: 2012-2014. Office for National Statistics, release date 23rd September 2015.
Woods R. Long-term trends in fetal mortality: implications for developing countries. Bulletin of the World Health Organization Volume 86 6;460-6.

9. Historical longevity

Andersen SL *et al*. Health span approximates life span among many supercentenarians: Compression of morbidity at the approximate limit of life span. J Gerentol A Biol Sci Med Sci 2012:67(A);395-405.
Department of Work and Pensions. Number of future centenarians by age group. April 2011.
Dong X *et al*. Evidence for a limit to human lifespan. Nature 2016:538;257-9.
Engberg H *et al*. Centenarians-a useful model for healthy aging? A 29-year follow-up of hospitalizations among 40,000 Danes born in 1905. Aging Cell 2009:8;270-6.
Fries JF *et al*. Compression of morbidity 1980-2011: A focused review of paradigms and progress. Journal of Aging Research 2011;1-10.
Galofré-Vilà G. Heights across the last 2000 years in England. No 151 Oxford University Economic and Social History Series.
Gerontology Research Group: http://www.grg.org. List of supercentenarians.
Griffin JP. Changing life expectancy throughout history. J R Soc Med 2008:101;577.
Milman S & Barzilai N. Dissecting the mechanisms underlying unusually successful human healthspan and lifespan. Cold Spring Harb Perspect Med 2016:6;a025098.
Montagu JD. Length of life in the ancient world: a controlled study. J R Soc Med 1994:87;25-6.
Quetelet 1842 quoted in Govindaraju D. Applied translational genomics 2015.
Statistical bulletin: Estimates of the very old (including centenarians): England and Wales, and United Kingdom, 2002 to 2014. Office for National Statistics, release date 30th September 2015.
Steenstrup T *et al*. Telomeres and the natural lifespan limit in humans. Aging 2017:9;1130-40.

Vaupel JW. Biodemography of human ageing. Nature 2010:464;536-42.

10. Ageing: Visible effects
Belsky DW *et al*. Quantification of biological aging in young adults. PNAS 2015:E4104-10.
Kennedy BK *et al*. Aging: a common driver of chronic diseases and a target for novel interventions. Cell 2014:159;709-13.
Liochev S. Which is the most significant cause of aging? Antioxidants 2015:4;793-810.
López-Otin C *et al*. The hallmarks of aging. Cell 2013:153;1194-1217.
Lustgarten MS. Classifying aging as a disease: The role of microbes. Frontiers in Genetics 2016:7;Article 212.
Melov S. Geroscience approaches to increase healthspan and slow aging. F1000Research 2016:59F1000 Faculty Rev;785.
Tosato M *et al*. The aging process and potential interventions to extend life expectancy. Clinical Interventions in Aging 2007:2;401-12.
Umbrello G & Esposito S. Microbiota and neurologic diseases: potential effects of probiotics. J Transl Med 2016:14;298-308.
Zhang Y *et al*. Hypothalamic stem cells control ageing speed partly through exosomal miRNAs. Nature 2017:online 26 July.

11. Ageing: Cellular effects
Atzmon G *et al*. Genetic variation in human telomerase is associated with telomere length in Ashkenazi centenarians. PNAS 2010:107;Suppl1;1710-7.
Baker DJ *et al*. Naturally occurring p 16Ink4a-positive cells shorten healthy lifespan. Nature 2016:530;184-9.
Bailey R. Evaluating calorie intake. Data Science Campus of the UK Office for National Statistics, February 2018 (https://datasciencecampus.ons.gov.uk/2018/02/15/eclipse/)
Bär C & Blasco MA. Telomeres and telomerase as therapeutic targets to prevent and treat age-related diseases. F1000Research 2016:5(F1000 Faculty Rev);89.
Bhatia-Dey N *et al*. Cellular senescence as the causal nexus of aging. Front Genet 2016:7;Article 13.
Bodnar AG *et al*. Extension to life-span by introduction of telomerase into normal human cells. Science 1998:279;349-52.
Callaghan MF *et al*. Widespread age-related differences in the human brain microstructure revealed by quantitative magnetic resonance imaging. Neurology of Aging 2014:35;1862-72.
Carmona JJ & Michan S. Biology of healthy aging and longevity. Rev Inves clin 2016:68;7-16.
Carnes BA *et al*. Does senescence give rise to disease? Mech Ageing Dev 2008:129;693-9.
Chandrasekaran A *et al*. Redox control of senescence and age-related disease. Redox Biology 2017:11;91-102.

Childs BG et al. Cellular senescence in aging and age-related disease: from mechanisms to therapy. Nat Med 2015:21;1424-35.
Chistiakov DA et al. Mitochondrial aging and age-related dysfunction of mitochondria. Biomed Res Int 2014:238463.
Doonan R *et al*. Against the oxidative damage theory of aging: superoxide dismutases protect against oxidative stress but have little or no effect on life span in *Caenorhabditis elegans*. Genes Dev 2008;22:3236-41.
Elmore S. Apoptosis: A review of programmed cell death. Toxicol Pathol 2007:35;495-516.
Glei DA *et al*. Predicting survival from telomere length versus conventional predictors: A multinational population-based cohort study. PLOS ONE 2016:11;e0152486.
Harris SE *et al*. Longitudinal telomere length shortening and cognitive and physical decline in later life: the Lothian birth cohorts 1936 and 1921. Mech Ageing Dev 2016:154;43-8.
Hornsby PJ. Telomerase and the aging process. Exp Gerontol 2007:42;575-81.
López-Otin C *et al*. The hallmarks of aging. Cell 2013:153;1194-1217.
Marioni RE *et al*. The epigenetic clock and telomere length are independently associated with chronological age and mortality. International Journal of Epidemiology 2016:45;424-32.
Shammas MA. Telomeres, lifestyle, cancer and aging. Curr Opin Clin Nutr Metab Care 2011:14;28-34.
Simons MJ. Questioning causal involvement of telomeres in aging. Ageing Research Reviews 2015:24;191-6.
Zhu H *et al*. Healthy aging and disease: role for telomere biology? CliSci (Lond) 2011:120(Pt 10);427-440.

12. Ageing: Genetics

Khan SS *et al*. A null mutation in SERPINE1 protects against biological aging in humans Sci Adv 2017;3:eaa01617.
Ayers E *et al*. Association of exceptional parental longevity and physical function in aging. AGE 2014:36;9677-83.
Beekman M *et al*. Genome-wide linkage analysis for human longevity: genetics of Healthy Aging Study. Aging Cell 2013:12:184-93.
Blagosklonny MV. Why human lifespan is rapidly increasing: solving "longevity riddle" with "revealed-slow-aging" hypothesis. Aging 2010:2;177-82.
Bocklandt S et *al*. Epigenetic predictor of age. PLOS One 2011:6;e14821.
Brooks-Wilson AR. Genetics of healthy aging and longevity. Hum Genet 2013:132;1323-8.
Budovsky A *et al*. LongevityMap: a database of human genetic variants associated with longevity. Trends in Genetics 2013:29;559-60.
Carrero D *et al*. Hallmarks of progeroid syndromes: lessons from mice and reprogrammed cells. Dis Model Mech 2016:9;719-35.
Fortney K *et al*. Genome-wide scan informed by age-related disease identifies loci

for exceptional human longevity. PLOS Genet 2015:11;e1005728.

Gordon LB *et al*. Clinical trial of the protein farnesylation inhibitors lonafarnib, pravastatin and zoledronic acid in children with Hutchinson-Gilford progeria syndrome. Circulation 2016:134;114-25.

Govindaraju D *et al*. Genetics, lifestyle and longevity: Lessons from centenarians. Applied and Translational Genomics 2015:4;23-32.

Hannum G *et al*. Genome-wide methylation profiles reveal quantitative views of human aging rates. Mol Cell 213:49;359-67.

Horvath S. DNA methylation age of human tissues and cell types. Genome Biology 2013:14;R115

Huidobro C *et al*. Aging epigenetics: causes and consequences. Mol Aspects Med 2013:34;765-81.

Johnson SC *et al*. Genetic evidence for common pathways in human age-related diseases. Aging Cell 2015:14;809-17.

Jones MJ *et al*. DNA methylation and healthy human aging. Aging Cell 2015:14;924-32.

Kenyon C. The plasticity of aging: Insights from long-lived mutants. Cell 2005:120;449-60.

Murabito JM *et al*. The search for longevity and healthy aging genes: insights from epidemiological studies and samples of long-lived individuals. J Gerontol A Biol Sci Med Sci 2012:67;470-9.

Newman AB & Murabito JM. The epidemiology of longevity and exceptional survival. Epidemiol Rev 2013:35;181-97.

Ostan R *et al*. Gender, aging and longevity in humans: an update of an intriguing/ neglected scenario paving the way to a gender-specific medicine. Clinical Science 2016:130;1711-25.

Passarinao G *et al*. Human longevity: Genetics or lifestyle? It takes two to tango. Immunity & Ageing 2016:13:12.

Pilling LC *et al*. Human longevity is influenced by many genetic variants: evidence from 75,000 UK Biobank participants. Aging 2016:8;547-60.

Rando TA & Chang HY. Aging, rejuvenation, and epigenetic reprogramming: resetting the aging clock. Cell 2012:148;46-57.

Reddy S & Comai L. recent advances in understanding the role of lamins in health and disease. F1000Research 2016:5(F1000 Faculty Rev);2536.

Scaffidi P & Misteli T. Reversal of the cellular phenotype in the premature aging disease Hutchinson-Gilford Progeria syndrome. Nat Med 2005:11;440-445.

Scaffidi P & Misteli T. Lamin A-dependent nuclear defects in human aging. Science 2006:312;1059-63.

Teschendorff AE *et al*. Age-associated epigenetic drift: implications, and a case of epigenetic thrift? Hum Mol Genet 2013:22(R1);R7-R15.

13. Ageing: Other organisms

Boehm A-M *et al*. FoxO is a critical regulator of stem cell maintenance in immortal

Hydra. PNAS 2012;109:19697-702

AnAge website: http://genomics.senescence.info/species/

Bartke A. Single-gene mutations and healthy ageing in mammals. Phil Trans R Soc B 2011:366;28-34.

Bilinski T. Principles of alternative gerontology. Aging 2016:8;589-602.

Clancy DJ *et al.* Extension of life-span by loss of CHICO, a Drosophila insulin receptor substrate protein. Science 2001:292;104-6.

Devarapalli P *et al.* The conserved mitochondrial gene distribution in relatives of *Turritopsis nutricula*, an immortal jellyfish. Bioinformation 2014;10:586-91.

Finch CE. Evolution of the human lifespan and diseases of aging: Roles of infection, inflammation and nutrition. PNAS 2010:107 suppl 1;1718-24.

Hashimoto T *et al.* Extremotolerant tardigrade genome and improved radiotolerance of human cultured cells by tardigrade-unique protein. Nat Commun 2016:7;12808.

Jones OR *et al.* Diversity of ageing across the tree of life. Nature 2014:505;169-173.

Kenyon C. The first long-lived mutants: discovery of the insulin/IGF-1 pathway for ageing. Phil Trans R Soc B 2011:366;9-16.

Kim EB *et al.* Genome sequencing reveals insights into physiology and longevity of the naked mole rat. Nature 2012:479;223-7.

MacRae SL *et al.* DNA repair in species with extreme lifespan differences. Aging 2015:7;1171-84.

MacRae SL *et al.* Comparative analysis of genome maintenance genes in naked mole rat, mouse and human. Aging Cell 2015:14;288-91.

Park TJ *et al.* Fructose-driven glycolysis supports anoxia resistance in the naked mole-rat. Science 2017:356;307-11.

Petralia RS *et al.* Aging and longevity in the simplest animals and the quest for immortality. Ageing Res Rev 2014:0;66-82.

Piersigilli A & Meyerholz DK. The "naked truth": Naked mole-rats do get cancer. Veterinary Pathology 2016:53;519-20.

Piraino S *et al.* Reversing the life cycle: Medusae transforming into polyps and cell transdifferentiation in Turritopsis nutricula (Cnidaria, Hydrozoa). Biol.Bull. 1996:190;302-12.

Ricklefs RE. Insights from comparative analyses of aging in birds and mammals. Aging Cell 2010:9;273-84.

Salguero-Gómez R *et al.* COMADRE: a global data base of animal demography. Journal of Animal Ecology 2016:85;371-84.

Schaible R *et al.* Constant mortality and fertility over age in Hydra. Proc Natl Acad Sci USA 2015:112;15701-6.

Uno M and Nishida E. Lifespan-regulating genes in *C. elegans*. NPJ Aging Mech Dis 2016:2:16010.

Vijg J & Campisi J. Puzzles, promises and a cure for ageing. Nature 2008:454;1065-71.

14. Lifespan: Effects of lifestyle

Antero-Jacquemin J da S *et al.* Learning from leaders: Life-span trends in Olympians and supercentenarians. J Gerontol A Biol Sci Med Sci 2015:70;944-9.
Aune D *et al.* BMI and all cause mortality: systematic review and non-linear dose-response meta-analysis of 230 cohort studies with 3.74 million deaths among 30.3 million participants. BMJ 2016:353;i2156.
Bell CL *et al.* Late-life factors associated with healthy aging in older men. J Am Geriatr Soc 2014:62;880-8.
Biswas A *et al.* Sedentary time and its association with risk for disease incidence, mortality, and hospitalization in adults. Ann Intern Med 2015:162;123-32.
Cappuccio FP *et al.* Sleep duration and all-cause mortality: A systematic review and meta-analysis of prospective studies. Sleep 2010:33;585-92.
Chakravarty EF *et al.* Reduced disability and mortality among aging runners. Arch Intern Med 2008:168;1638-46.
Chakravarty EF *et al.* Lifestyle risk factors predict disability and death in healthy aging adults. Am J Med 2012:125;190-7.
Da Silva AA *et al.* Sleep duration and mortality in the elderly: a systematic review with meta-analysis. BMJ Open 2016:6;e008119.
Demakakos P *et al.* Wealth and mortality at older ages: a prospective cohort study. J Epidemiol Community Health 2016:70;346-53.
Emerging Risk Factors Collaboration. Association of cardiometabolic multimorbidity with mortality. JAMA 2015:314;52-60.
Flegal KM *et al.* Cause-specific excess deaths associated with underweight, overweight, and obesity. JAMA 2007:298;2028-37.
He Y *et al.* The transcriptional repressor DEC2 regulates sleep length in mammals. Science 2009:325;866-70.
Jha P *et al.* 21st-century hazards of smoking and benefits of cessation in the United States. N Engl J Med 2013:368;341-50.
Jones CR *et al.* Genetic basis of human circadian rhythm disorders. Exp Neurol 2013:243;28-33.
Khaw K-T *et al.* Combined impact of health behaviours and mortality in men and women: The EPIC-Norfolk prospective population study. PLOSMed 2008:5;e12.
Kitahara CM *et al.* Association between Class III obesity (BMI of 40-59 kg/m^2) and mortality: a pooled analysis of 20 prospective studies. PLOSMed 2014:11;e1001673.
Kyrgiou M *et al.* Adiposity and cancer at major anatomical sites: umbrella review of the literature. BMJ 2017:356;j477.
McNally S *et al.* Focus on physical activity can help avoid unnecessary social care. BMJ 2017;359;j4609.
Mossakowska M *et al.* How often and how much have Polish centenarians smoked? (Article in Polish). Przegl Lek 2006:63;1105-7.
Ngandu T *et al.* A 2 year multidomain intervention of diet, exercise, cognitive training, and vascular risk monitoring versus control to prevent cognitive

decline in at-risk elderly people (FINGER): a randomised controlled trial. Lancet 2015:385;2255-63.

Pase MP *et al.* Sugar- and artificially sweetened beverages and the risks of incident stroke and dementia. Stroke 2017:48;1139-46.

Pase MP *et al.* Sugary beverage intake and preclinical Alzheimer's disease in the community. Alzheimers Dement 2017:S1552-5260. (Epub ahead of print).

Rajpathak SN *et al.* Lifestyle factors of people with exceptional longevity. J Am Geriatr Soc 2011:59;1509-12.

Reimers CD *et al.* Does physical activity increase life expectancy? Journal of Aging Research 2012:243958.

Ri M *et al.* Effects of body mass index (BMI) on surgical outcomes: A nationwide survey using a Japanese web-based database. Surg Today 2015:45;1271-9.

Robinson M *et al.* Enhanced protein translation underlies improved metabolic and physical adaptations to different exercise training modes in young and old humans. Cell Metabolism 2017:25;581-92.

Smyth A *et al.* Alcohol consumption and cardiovascular disease, cancer, injury, admission to hospital, and mortality. Lancet 2015:386;1945-54.

Wei M *et al.* Fasting – mimicking diet and markers/risk factors for aging, diabetes, cancer and cardiovascular disease. Science Translational Medicine 2017:377;eaai8700.

Wilhelmsen L *et al.* Factors associated with reaching 90 years of age: a study of men born in 1913 in Gothenburg, Sweden. J Intern Med 2011:269;441-51.

Yates LB *et al.* Exceptional longevity in men. Arch Intern Med 2008:168;284-290.

15. Lifespan: Maintenance and repair

Abdul-Ghani M *et al.* Cardiotrophin 1 stimulates beneficial myogenic and vascular remodeling of the heart. Cell Research 2017:advanced online 8 August 2017.

Ali MA *et al.* Metabolic and immune effects of immunotherapy with proinsulin peptide in human new-onset type 1 diabetes. Science Translational Medicine 2017:402;eaaf7779.

16. Lifespan: Nutrition

Andrieu S *et al.* Effect of long-term omega 3 polyunsaturated fatty acid supplementation with or without multidomain intervention on cognitive function in elderly adults with memory complaints (MAPT): a randomised, placebo-controlled trial. Lancet Neurol 2017:May 16(5);377-89.

Aune D *et al.* Fruit and vegetable intake and the risk of cardiovascular disease, total cancer and all-cause mortality: a systematic review and dose-response meta-analysis of prospective studies. Int J Epidemiol 2017:(Epub ahead of print).

Bolland MJ *et al.* Calcium intake and risk of fracture: systematic review. BMJ 2015:351;h4580.

Bolland MJ *et al.* Should adults take vitamin D supplements to prevent disease? BMJ 2016:355;i6201.

Bonaccio M *et al.* High adherence to the Mediterranean diet is associated with cardiovascular protection in higher but not in lower socioeconomic groups: prospective findings from the Moli-san study. Int J Epidemiol 2017:dyx145

Bjelakovic G *et al*. Antioxidant supplements for prevention of mortality in healthy participants and patients with various diseases. Cochrane Database Syst Rev 2012:3;CD007176.

Conner TS *et al*. The role of personality traits in young adult fruit and vegetable consumption. Front Psychol 2017:8:Article119.

Coussens AK *et al*. Vitamin D accelerates resolution of inflammatory responses during tuberculosis treatment. Proc Natl Acad Sci U S A 2012:109;15449-54.

Dehghan *et al*. Associations of fats and carbohydrate intake with cardiovascular disease and mortality in 18 countries from five continents (PURE): a prospective cohort study. Lancet 2017:S0140-6736(17)32252-3.

Douaud G. *et al*. Preventing Alzheimer's disease-related gray matter atrophy by B-vitamin treatment. Proc Natl Acad Sci U S A, 2013:110;9523-8.

Dubé C *et al*. The prevalence of celiac diseases in average-risk and at-risk Western European populations: a systematic review. Gatroenterology 2005:128(4 Suppl 1);S57-67.

Fiolet T *et al*. Consumption of ultra-processed foods and cancer risk: results from NutriNet-Santé prospective cohort. BMJ 2018:360;k322.

Freedhoff Y. From plunger to Punkt-roller: a century of weight-loss quackery. CMAJ 2009:180;432-3.

Guasch-Ferré M *et al*. Nut consumption and risk of cardiovascular disease J Am Coll Cardiol 2017;70:2519-32.

Knoops KT *et al*. Mediterranean diet, lifestyle factors, and 10-year mortality in elderly European men and women: the HALE project. JAMA 2004:292;1433-9.

Lebwohl B *et al*. Long term gluten consumption in adults without celiac disease and risk of coronary heart disease: prospective cohort study. BMJ 2017:357;j1892.

Martineau AR *et al*. Vitamin D supplementation to prevent acute respiratory tract infections: systematic review and meta-analysis of individual participant data. BMJ 2017:356;i6583.

Micha R *et al*. Association between dietary factors and mortality from heart disease, stroke, and type 2 diabetes in the United States. JAMA 2017:317;912-24.

Miller V *et al*. Fruit, vegetable and legume intake, and cardiovascular disease and deaths in 18 countries (PURE): a prospective cohort study. Lancet 2017:S0140-6736(17)32253-5.

Morgan Spurlock website: http://morganspurlock.com/work/super-size-me/

Mueller NT & Appel LJ. Attributing death to diet precision counts. JAMA 2017:317;908-9.

Pedro de Magãlhes J *et al*. Fish oil supplements, longevity and aging. Aging 2016:8;1578-82.

Preiser J-C *et al*. Metabolic and nutritional support of critically ill patients: consensus and controversies. Critical Care 2015:19;35-45.

Robinson SM *et al*. Adult lifetime diet quality and physical performance in older

age: Findings from a British birth cohort. J Gerontol A Biol Sci Med Sci 2017; online access October 2017.
Shi H *et al*. NAD deficiency, congenital malformations, and niacin supplementation. N Engl J Med 2017:377;544-52.
Sydenham E *et al*. Omega 3 fatty acid for the prevention of cognitive decline and dementia. Cochrane Database Syst Rev. 2012 Jun 13;(6)
Zhong J *et al*. B vitamins attenuate the epigenetic effects of ambient fine particles in a pilot human intervention trial. Proc Natl Acad Sci U S A 2017:114;3503-8.
NHS portion size advice: www.nhs.uk/Livewell/5ADAY/Pages/Portionsizes.aspx
Information about vitamins:
www.nhs.uk/Conditions/vitamins-minerals/Pages/vitamins-minerals.aspx
For a review of dietary supplements see:
Supplements Who needs them? A Behind the Headlines report June 2011.
NHS choices www.nhs.uk/news/2011/05May/Documents/BtH_supplements.pdf
GAO (United States Government Accountability Office_ Report to Congressional Requesters. Dietary Supplements.
www.gao.gov/assets/660/653113.pdf
www.gao.gov/new.items/d011129.pdf
Health products for seniors. Potential harm from 'Anti-Aging' products. GAO (United States Government Accountability Office. Testimony before the special committee on aging, U.S. Senate. Released September 2001.

17. *Extending life: Introduction*

Kaeberlein M et al. Healthy aging: the ultimate preventative medicine. Science 2015:350;1191-3.
Kenyon C. The plasticity of aging: insights from long-lived mutants. Cell 2005:120;449-60.
Longo VD *et al*. Interventions to slow aging in humans: are we ready? Aging Cell 2015:14;497-510.
Pedro de Magalhães J. The scientific quest for lasting youth: prospects for curing aging. Rejuvenation Research 2014:17;458-67.
Rattan SIS. Aging is not a disease: Implications for intervention. Aging and Disease 2014:5;196-202.
Vijg J & de Grey ADNJ. Innovating aging: promises and pitfalls on the road to life extension. Gerontology 2014:60;373-80.
Vijg J. and Kennedy BK. The essence of aging. Gerontology 2016:62;381-5.

18. *Extending life: Evolution*

Bellis MA *et al*. Measuring paternal discrepancy and its public health consequences. J Epidemiol Community Health 2005:59;749-54.
Baudisch A and Vaupel JW. Getting to the root of aging. Science 2012:338;618-9.
Darwin awards: http://darwinawards.com
Wilder SM & Rypstra AL. Sexual size dimorphism predicts the frequency of sexual cannibalism within and among species of spiders. Am Nat 2008:172;431-40.

20. Extending life: Maintenance and repair

Abbosh C *et al*. Phylogenetic ctDNA analysis depicts early stage lung cancer evolution. Nature 2017: online 26th April.

Ajiboye A *et al*. Restoration of reaching and grasping movements through brain-controlled muscle stimulation in a person with tetraplegia. Lancet 2017:S0140-6736 Epub ahead of print.

Butler C.R *et al*. Vacuum-assisted decellularization: an accelerated protocol to generate tissue-engineered human tracheal scaffolds. Biomaterials 2017:124;95-105.

Famm K *et al*. Drug discovery: A jump-start for electroceuticals. Nature 2013:496;159-61.

Grau C *et al*. Conscious brain-to-brain communication in humans using non-invasive technologies. PLOS One 2014:9;e105225.

Kahl LJ & Endy D. A survey of enabling technologies in synthetic biology. Journal of Biological Engineering 2013:7;13-30.

Kumar S & Chatterjee K. Comprehensive review on the use of graphene-based substrates for regenerative medicine and biomedical devices. ACS Appl Mater Interfaces 2016:8;26431-57.

Lee J. Cochlear implantation, enhancements, transhumanism and posthumanism. Some human questions. Sci Eng Ethics 2016:22;67-92.

Leuhardt EC *et al*. The emerging world of motor neuroprosthetics: a neurosurgical perspective. Neurosurgery 2006:59;1-14.

Li X *et al*. Digital health: Tracking physiomes and activity using wearable biosensors reveals useful health-related information. PLOS Biol 2017:15;e2001402.

McNamee MJ & Edwards SD. Transhumanism, medical technology and slippery slopes. J Med Ethics 2006:32;513-8.

Naderi H *et al*. Review paper: critical issues in tissue engineering: biomaterials, cell sources, angiogenesis, and drug delivery systems. J Biomater Appl 2011:26;383-417.

Niu D *et al*. Inactivation of porcine endogenous retrovirus in pigs using CRISPR-Cas9. Science 2017:eaan4187.

Orlando G. *et al*. Regenerative medicine as applied to general surgery. Ann Surg 2012:255;867-80.

Pais-Vieira M *et al*. Building an organic computing device with multiple interconnected brains. Sci Rep 2015:5;11869.

Phallen J *et al*. Direct detection of early-stage cancers using circulating tumour DNA. Science Translational Medicine 2017:403;eaan2415.

Wertheim JA *et al*. Cellular therapy and bioartificial approaches to liver replacement. Curr Opin Organ Transplant 2012:17;235-40.

Wu G-H & Hsu S-h. Polymeric-based 3D printing for tissue engineering. J Med Biol Eng 2015:35;285-92.

Wu J *et al*. Generation of human organs in pigs via interspecies blastocyst complementation. Reprod Domest Anim 2016:Suppl 2;18-24.

Zehr EP. Future think: cautiously optimistic about brain augmentation using tissue engineering and machine interface. Front Syst Neurosci 2015:9;Article 72.

21. Extending life: Calorie or dietary restriction

Cheng C-W *et al*. Fasting-mimicking diet promotes Ngn3-driven ß-cell regeneration to reverse diabetes. Cell 2017:168;775-8.

Colman RJ *et al*. Caloric restriction reduces age-related and all-cause mortality in rhesus monkeys. Nat Commun 2014:5;3557.

Fontana L *et al*. Calorie restriction or exercise: effect on coronary heart disease risk factors. Am J Physiol Endocrinol Metab 2007:293;E197-202.

Grandison RC *et al*. Amino acid imbalance explains extension of lifespan by dietary restriction in Drosophila. Nature 2009;462:1061-4.

Guarente L & Picard F. Calorie restriction-the SIR2 connection. Cell 2005:120;473-82.

Gillespie ZE *et al*. Better living through chemistry: Caloric restriction (CR) and CR mimetics alter genome function to promote increased health and lifespan. Frontiers in Genetics 2016:7;Article 142.

Heilbronn LK et al. Effect of 6-month calorie restriction on biomarkers of longevity, metabolic adaptation, and oxidative stress in overweight individuals. JAMA 2006:295;1539-48.

Kemnitz JW. Calorie restriction and aging in nonhuman primates. ILAR J 2011:52;66-77.

Lee S-G *et al*. Age-associated molecular changes are deleterious and may modulate life span through diet. Sci Adv 2017:3;e1601833.

López-Lluch G & Navas P. Calorie restriction as an intervention in ageing. J Physiol 2016:594;2043-60.

Mattison JA *et al*. Impact of caloric restriction on health and survival in rhesus monkeys from the NIA study. Nature 2012:489;318-21.

Mattison JA *et al*. Caloric restriction improves health and survival of rhesus monkeys. Nat Commun 2017:8;14063.

McDonald RB & Ramsey JJ. Honoring Clive McCay and 75 years of calorie restriction research. J Nutr 2010:140;1205-10.

Ravussin E *et al*. A 2-year randomized controlled trial of human caloric restriction: feasibility and effects on predictors of health span and longevity. J Gerontol A Biol Sci Med Sci 2015:70;1097-1104.

Redman LM *et al*. Effect of calorie restriction with or without exercise on body composition and fat distribution. J Clin Endocrinol Metab 2007:92;865-72.

Sohal RS & Forster MJ. Caloric restriction and the aging process: Free Radic Biol Med 2014:0;366-82.

Weindruch R & Walford RL. Dietary restriction in mice beginning at 1 year of age: effect on life-span and spontaneous cancer incidence. Science 1982:215;1415-8.

22. Extending life: Drugs and nootropics

Barzilai N *et al*. Metformin as a tool to target aging. Cell Metabolism 2016:23;1060-5.

Longevity

Calabrese V *et al.* Cellular stress response: a novel target for chemoprevention and nutritional neuroprotection in aging, neurodegenerative disorders and longevity. Neurochem Res 2008:33;2444-71.
Database of geroprotectors:.Geroprotectors.org
Ehninger D *et al.* Longevity, aging and rapamycin. Cell Mol Life Sci 2014:71;4325-46.
Flicker L & Grimley Evans G. Piracetam for dementia or cognitive impairment. Cochrane Database Syst Rev 2001:(2);CD001011.
Imai S-I & Guarente L. It takes two to tango: NAD+ and sirtuins in aging/longevity control. Aging and Mechanisms of Disease 2016:2;16017.
Mallikarjun V & Swift J. Therapeutic manipulation of ageing: repurposing old dogs and discovering new tricks. EBioMedicine 2016:14;24-31.
Miller RA *et al.* Rapamycin-mediated lifespan increase in mice is dose and sex dependent and metabolically distinct from dietary restriction. Aging Cell 2014:13;468-77.
Mills KF *et al.* Long-term administration of nicotinamide mononucleotide mitigates age-associated physiological decline in mice. Cell Metab 2016:24;795-806.
Moskalev A *et al.* Developing criteria for evaluation of geroprotectors as a key stage towards translation to the clinic. Aging Cell 2016:15;407-15.
Newman JC *et al.* Strategies and challenges in clinical trials targeting human aging. J Gerontol A Biol Sci Med Sci 2016:71;1424-34.
Nobre AC *et al.* L-theanine, a natural constituent in tea, and its effect on mental state. Asia Pac J Clin Nutr 2008:17 Suppl 1;167-8.
Nootropics website: https://nootropics.com
Schmitt R. Senotherapy: growing old and staying young? Pflugers Arch. 2017;Apr 7 epub.
Semba RD *et al.* Resveratrol levels and all-cause mortality in older community-dwelling adults. JAMA Intern Med 2014:174;1077-84.
Testa G *et al.* Calorie restriction and dietary restriction mimetics: a strategy for improving healthy aging and longevity. Curr Pharm Des 2014:20;2950-77.
Timmers S *et al.* Calorie restriction-like effects of 30 days of resveratrol supplementation on energy metabolism and metabolic profile in obese humans. Cell Metab 2011:14;612-22.
Vaiserman A and Lushchak O. Implementation of longevity-promoting supplements and medications in public health practice: achievements, challenges and future perspectives. J Transl Med 2017;15:160-9.
Vaiserman AM & Pasyukova EG. Epigenetic drugs: a novel anti-aging strategy? Frontiers in Genetics 2012:3;article 224.

23. Extending life: Gene therapy

Bernades de Jesus B *et al.* Telomerase gene therapy in adult and old mice delays aging and increases longevity without increasing cancer. EMBO Mol Med 2012:4;691-704.
Bradley RW *et al.* Tools and principles for microbial gene circuit engineering. J Mol

Biol 2016:428;862-88.

Bourret R *et al*. Human-animal chimeras: ethical issues about farming chimeric animals bearing human organs. Stem Cell Res Ther 2016:7;87-93.

ENCODE Project Consortium. Identification and analysis of functional elements in 1 per cent of the human genome by the ENCODE pilot project. Nature 2007:447;799-816.

GeneAge database: http://genomics.senescence.info/genes/

Graur D *et al*. On the immortality of television sets: "Function" in the human genome according to the evolution-free gospel of ENCODE. Genome Biol. Evol. 2013:5;578-90.

Kellis M *et al*. Defining functional DNA elements in the human genome. PNAS 2014:111;6131-38.

Kumar SRP *et al*. Clinical development of gene therapy: results and lessons from recent successes. Mol Ther Methods Clin Dev 2016:3;16034.

Kungulovski G & Jeltsch A. Epigenome editing: State of the art, concepts and perspectives. Trends Genet 2016:32;101-13.

Laufer BI & Singh SM. Strategies for precision modulation of gene expression by epigenome editing: an overview. Epigenetics & Chromatin 2015:8;34-45.

Ma H *et al*. Correction of a pathogenic gene mutation in human embryos. Nature 2017:548;413-9.

Martella A *et al*. Mammalian synthetic biology: Time for big MACs. ACS Synth Biol 2016:5;1040-9.

Nathwani AC *et al*. Long-term safety and efficacy of Factor IX gene therapy in hemophilia B. NEJM 2014:371;1994-2004.

Nemudryi AA *et al*. TALEN and CRISPR/Cas genome editing systems: tools of discovery. Acta Naturae 2014:6;19-40.

Nielsen AA *et al*. Genetic circuit design automation. Science 2016:352;aac7341.

Ribeil J-A *et al*. Gene therapy in a patient with sickle cell disease. N Engl J Med 2017:376;848-55.

Rice MK & Ruder WC. Creating biological nanomaterials using synthetic biology. Sci Technol Adv Mater 2014:15(1).

Richardson SM *et al*. Design of a synthetic yeast genome. Science 2017:355;1040-4.

Rogers JK. & Church GM. Multiplexed engineering in biology. Trends Biotechnol 2016:34;198-206.

Sampson TR & Weiss DS. Exploiting CRISPR/Cas systems for biotechnology. Bioessays 2014:36;34-8.

Tang L *et al*. CRISPR/Cas9-mediated gene editing in human zygotes using Cas9 protein. Mol Genet Genomics 2017: March (Epub)

Vassena *et al*. Genome engineering through CRISPR/Cas9 technology in the human germline and pluripotent stem cells. Hum Reprod Update 2016;22;411-9.

Xu T *et al*. Cas9-based tools for targeted genome editing and transcriptional control. Appl Environ Microbiol 2014:80;1544-52.

24. Extending life: Regeneration

Bitto A & Kaeberlein M. Rejuvenation: it's in our blood. Cell Metab 2014:20;2-4.

Conboy IM *et al*. Rejuvenation of aged progenitor cells by exposure to a young systemic environment. Nature 2005:433;760-4.

Conboy IM *et al*. Systemic problems: a perspective on stem cell aging and rejuvenation. Aging 2015:7;754-65.

Conboy MJ *et al*. Heterochronic parabiosis: historical perspective and methodological considerations for studies of aging and longevity. Aging Cell 2013:12;525-30.

Garcia-Dominguez X *et al*. First steps towards organ banks: vitrification of renal primordial. Cryo Letters 2016:37;47-52.

Harrison SE *et al*. Assembly of embryonic and extraembryonic stem cells to mimic embryogenesis *in vitro*. Science 2017:356 (Epub).

Ji P *et al*. Induced pluripotent stem cells: Generation strategy and epigenetic mystery behind reprogramming. Stem Cells Int 2016:8415010.

Kalladka D *et al*. Human neural stem cells in patients with chronic ischaemic stroke (PISCES): a phase 1, first-in-man study. Lancet 2016:388;787-96.

Kikuchi T *et al*. Human iPS cell-derived dopaminergic neurons function in a primate Parkinson's disease model. Nature 2017:548;592-6.

Kimbrel EA & Lanza R. Pluripotent stem cells: the last 10 years. Regen Med 2016:11;831-47.

Ma T *et al*. Progress in the reprogramming of somatic cells. Circ Res 2013:112;562-74.

Manukyan M & Singh PB. Epigenetic rejuvenation. Genes to Cells 2012:17;337-43.

Marión RM & Blasco MA. Telomere rejuvenation during nuclear reprogramming. Curr Opin Genet Dev 2010:20;190-6.

Mason JO & Price DJ. Building brains in a dish: Prospects for growing cerebral organoids from stem cells. Neuroscience 2016:334;105-18.

Ocampo A. *et al*. *In vivo* amelioration of age-associated hallmarks by partial reprogramming. Cell 2016:167;1719-33.

Palomo ABA *et al*. The power and the promise of cell reprogramming: personalized autologous body organ and cell transplantation. J Clin Med 20143:373-87.

Rebo J *et al*. A single heterochronic blood exchange reveals rapid inhibition of multiple tissues by old blood. Nature Communications 2016:13363.

Rohani L *et al*. The aging signature: a hallmark of induced pluripotent stem cells? Aging Cell 2014:13;2-7.

Shieh S-J & Cheng T-C. Regeneration and repair of human digits and limbs: fact and fiction. Regeneration 2015:13;149-68.

Singh S *et al*. Induced pluripotent stem cells: applications in regenerative medicine, disease modeling, and drug discovery. Front Mol Biosci 2015:3;Article 2.

Singh VK *et al*. Advances in stem cell research-A ray of hope in better diagnosis and prognosis in neurodegenerative diseases. Front Mol Biosci 2016:3;Article 72.

Trakarnsanga K *et al*. An immortalized adult human erythroid line facilitates sustainable and scalable generation of functional red cells. Nature

Communications 2017:8;14750.

Windrem MS *et al*. A competitive advantage by neonatally engrafted human glial progenitors yields mice whose brains are chimeric for human glia. J Neurosci 2014:34;16153-61.

Wu J & Belmonte JCI. Dynamic pluripotent stem cell states and their applications. Cell Stem Cell 2015:17;509-25.

Wu J *et al*. Stem cells and interspecies chimaeras. Nature 2016:540;51-9.

Wu J. Interspecies chimerism with mammalian pluripotent stem cells. Cell 2017:168;473-86.

Wyss-Coray T Ageing, neurodegeneration and brain rejuvenation. Nature 2016:539;180-6.

Yee J Turning somatic cells into pluripotent stem cells. Nature Education 2010:3;25-9.

Yousefi A-M *et al*. Prospect of stem cells in bone tissue engineering: A review. Stem Cells Int 2016:epub Jan 2016.

Yun MH. Changes in regenerative capacity through lifespan. Int J Mol Sci 2015:16;25392-432.

Zhao W *et al*. Regulatory factors of induced pluripotency: current status. Stem Cell Investig 2014:1;15-26.

25. Extending life: Cryronics and hibernation

Blanco MB *et al*. Hibernation in a primate: does sleep occur? R Soc Open Sci 2016:3;160282.

Cerri M. The central control of energy expenditure: Exploiting torpor for medical applications. Annual Review of Physiology 2017:79;167-186.

Faherty SL *et al*. Gene expression profiling in the hibernating primate, Cheirogaleus Medius. Genome Biol Evol 2016:8;2413-26.

Storey KB. Out cold: Biochemical regulation of mammalian hibernation. Gerontology 2010:56;220-30.

Vita-More N & Barranco D. Persistence of long-term memory in vitrified and revived *Caenorhabditis elegans*. Rejuvenation Research 2015:18;458-63.

Zhao S *et al*. Genomic analysis of expressed sequence tags in American black bear Ursus americanus. BMC Genomics 2010:11;201-16.

Also see: www.fda.gov/ucm/groups/fdagov-public/@fdagov-afda-adcom/ documents/document/ucm367601.pdf] and a Cochrane review from 2014: www. cochrane.org/CD007260/VASC_mechanical-chest-compression-machines-for-cardiac-arrest.

26. Extending life: Uploading

Cerullo MA. Uploading and branching identity. Minds and Machines 2015:25;17-36.

Draganski B *et al*. Computational anatomy for studying use-dependent brain plasticity. Frontiers in Human Neuroscience 2014:8;Article 380.

Eliasmith C *et al*. A large-scale model of the functioning brain. Science 2012:338;1202-5.

Koene RA. Fundamentals of whole brain emulation: state transition and update representations. International Journal of Machine Consciousness 2012:4;1250002.
Makram H *et al*. Reconstruction and simulation of neocortical microcircuitry. Cell 2015:163;456-92.
Sandberg A (2013). Feasibility of whole brain emulation. In Vincent C Müller (ed) Theory and Philosophy of Artificial Intelligence (SAPERE: Berlin: Springer), 251-64.

27. Extending life: Head transplant

Canavero S. The "Gemini" spinal cord fusion protocol: Reloaded. Surg Neurol Int 2015:6;18-25.
Lamba N *et al*. The history of head transplantation. Acta Neurochir 2016:158;2239-47.Li PW *et al*. A cross-circulated bicephalic model of head transplantation. CNS Neurosci Ther 2017: (Epub).

28. A world of millenarians

Ericsson KA *et al*. The role of deliberate practice in the acquisition of expert performance. Psychological Review 1993:100;363-406.
Jorde LB & Wooding SP. Genetic variation, classification and 'race'. Nat Genet 2004:36(11 Suppl);S28-33.
Lynch M. Mutation and human exceptionalism: Our future genetic load. Genetics 2016:202;869-75.

About the author

DAVID GOLDHILL is a retired intensive care doctor. His first degree was in politics and economics at Oxford University. He then qualified as a doctor at the London Hospital. His specialist training in anaesthesia and intensive care medicine took place in hospitals in the United Kingdom and the United States. Appointed an NHS consultant 1989 he worked at the Royal London Hospital and the Royal National Orthopaedic Hospital. He also had research and academic commitments associated with Colleges of the University of London. During his working life he was able to contribute to a variety of advisory bodies, committees and editorial boards, and participate in training and examining for medical students and doctors. His clinical responsibilities involved looking after the sickest patients admitted to hospital. Since retirement he's become fascinated by the ageing process and the growing body of evidence revealing the underlying mechanisms. He has an index linked NHS pension and the prospect of many years of productive life ahead of him. He is lucky enough to spend several months of the year sailing and skiing and helping to look after his grandchildren. Writing this book has been a fascinating way of using much of the rest of his time. Should he have taken up golf instead? Comments, corrections and suggestions are welcome at **livingto1000@gmail.com**

David Goldhill MA, MBBS, FRCA, FFICM, EDICM, MD.